Celia A Pollitz
Brown Univ   1982

THE PERCEPTION
OF ODORS

This is a volume in
**ACADEMIC PRESS
SERIES IN COGNITION AND PERCEPTION**

A Series of Monographs and Treatises

A complete list of titles in this series appears at the end of this volume.

THE PERCEPTION OF ODORS

Trygg Engen

Walter S. Hunter Laboratory of Psychology
Brown University
Providence, Rhode Island

ACADEMIC PRESS   1982
A Subsidiary of Harcourt Brace Jovanovich, Publishers
New York   London
Paris   San Diego   San Francisco   São Paulo   Sydney   Tokyo   Toronto

COPYRIGHT © 1982, BY ACADEMIC PRESS, INC.
ALL RIGHTS RESERVED.
NO PART OF THIS PUBLICATION MAY BE REPRODUCED OR
TRANSMITTED IN ANY FORM OR BY ANY MEANS, ELECTRONIC
OR MECHANICAL, INCLUDING PHOTOCOPY, RECORDING, OR ANY
INFORMATION STORAGE AND RETRIEVAL SYSTEM, WITHOUT
PERMISSION IN WRITING FROM THE PUBLISHER.

ACADEMIC PRESS, INC.
111 Fifth Avenue, New York, New York 10003

United Kingdom Edition published by
ACADEMIC PRESS, INC. (LONDON) LTD.
24/28 Oval Road, London NW1 7DX

Library of Congress Cataloging in Publication Data

Engen, Trygg, Date.
  The perception of odors.

  (AP series in cognition and perception)
  Bibliography: p.
  Includes index.
  1. Smell. 2. Smell--Psychological aspects.
3. Psychophysics. I. Title. II. Series.
QP458.E53   152.1'66   82-1653
ISBN 0-12-239350-3   AACR2

PRINTED IN THE UNITED STATES OF AMERICA

82 83 84 85   9 8 7 6 5 4 3 2 1

CONTENTS

Preface ix
Acknowledgments xi

1 INTRODUCTION 1

Odor Acuity 3
What Is the Odor Stimulus? 6
Odor Classification 7
Odor Mixture 9
Odor Coding 10
Hedonics 11
Odor Memory 14
Organization of This Book 15

2 GROSS ANATOMY AND PHYSIOLOGY 17

The Nasal Passages 17
The Epithelium 19
The Olfactory Receptors 20
The Olfactory Bulb 25
The Olfactory Brain 27
Coding 29

3 PSYCHOPHYSICS 35

Olfactometry 36
Odor-Control Technology 42
Odor Classification 44
Absolute Threshold and Detection 51
Difference Threshold and Resolving Power 54
Suprathreshold Odors: Psychophysical Scaling 56
Odor-Information Transmission 59

4 ADAPTATION 61

Definition of the Problem 61
Self-Adaptation 63
Cross-Adaptation 74

5 ANOSMIA 79

Heredity 80
The Classification of Disorders of Odor Perception 86
Aging 90
Hyperosmia 92
Effects of Anosmia 93

6 MEMORY 97

Identification of Odors 98
Odor Experts 101
The Influence of Verbal Labels on Odor Memory 102
Odor Memory without Verbal Labels 106
Factors That Influence Odor Memory 108
Forgetting 110
Is There Odor Recall? 111

7 ODOR MIXTURES 113

Perceived Intensity of Odor Mixtures 114
Perceived Quality of Odor Mixtures 116
The Nature of Masking 118

Psychophysical Theory 120
Is the Sense of Smell Analytic? 123

8
ODOR HEDONICS 125

Problems in Hedonic Measurement 126
Origin of Odor Preferences 130
Are There Universally Pleasant and Unpleasant Odors? 134
Do Odors Affect Health? 137
The Pleasantness of Odor and Bodily State 138
Is There a Human Pheromone? 139

9
INTERACTIONS OF ODOR AND OTHER PERCEPTIONS 145

Odor and Taste 145
Odor and Irritation 149
Odor and Light and Sound 153
The Unity of the Senses 155

10
PRACTICAL PROBLEMS AND POTENTIALS OF ODOR PERCEPTION 157

Odor Illusions 157
False Alarms 158
Perceptual Constancy 159
Applications of Odor Research 161

11
EPILOGUE 169

Incidental Associations 169
Odor Memory 170
Habituation and Arousal 171
The Coding of Odor Perception 172

References 173
Subject Index 197

PREFACE

Olfaction and the study of odor perception are usually only briefly discussed in textbooks on the senses, perception, and information processing. Audition and vision are accorded the most in-depth treatment. Yet some studies of olfaction indicate that there may be interesting differences and similarities between olfaction and the other senses. Information on odor perception provides an entirely different perspective on the general study of perception than does the study of any of the other senses—especially in the areas of affective reaction.

What is written about olfaction is often old and out-of-date. For example, Henning's odor prism is still presented, but often without mention of its detractors—including notable contemporaries of Henning—or of Henning's practical or theoretical bases. Largely ignored is a whole new literature on the perception of odors, which stems partly from Stevens' "new psychophysics" and signal detection theory and partly from the influence of cognitive psychology. This literature deserves attention by those who study the basic questions of perception and by those whose interests are more directly in the area of olfaction.

This book focuses attention on some of the new literature in a succinct, understandable way. It also covers some of the more exotic aspects of the study of olfaction, such as the possibility of human pheromones and the existence of odor pollution. Chapter 1 introduces the subject of olfaction, briefly tracing the history of the field and pointing out most of the significant historical markers along the way. Chapter 2 is a very general discussion of the anatomy and physiology of the olfactory system. Chapter 3, "Psychophysics," presents not only a history of the measurement of odors but also some of the new developments in the psychophysics of smell that have evolved from signal detection theory. The chapter also discusses some of the practical problems relating to the psychophysics of olfaction, for example, odor-control technology for industry. Chapters 4 through 8 are more specialized treatments of specific subject areas of olfaction. They are organized around currently active areas of research, such as odor masking, adaptation, deodorization, odor mixture, and memory. Chapter 9 relates the problems and findings on the perception of odor to the other senses. And Chapter 10 discusses the future of odor perception research both from the standpoint of practical problems to be overcome and potential developments in the field and their implications. Finally, Chapter 11 sums up the research discussed in the book and briefly mentions some of the more basic research areas, including the problem of the coding of odor perception.

It is safe to say that this book is like no other on the topic. Although it maintains a historical as well as an interdisciplinary perspective, its focus is on contemporary research with human subjects in sensory, perceptual, and psychophysical tasks. It attempts to deal with what is characteristic of odor perception psychologically rather than anatomically or physiologically. It is based on data rather than theory.

Except for some very short popular accounts, no books on odor perception have been published during the last decade. In light of the rather significant advances—and the fact that the number of empirically based papers published about olfaction may finally have exceeded the number of theoretical ones—it seems more important than ever to fill the need for a broad but cohesive treatment of the subject. Certainly, James' (1893) assertion in *Psychology* that smell and other chemical senses "need not be touched upon in this book, as almost nothing of psychological interest is known of them [p. 69]" is no longer valid.

This book will be of use as a supplement in undergraduate courses on sensation and perception. It will also be of use in graduate seminars and proseminars on more specific topics in the chemical senses, for which there is a great lack of the kind of reading material provided by this book. It should also be of interest to perfumers, medical investigators, and, of course, to psychologists and physiologists interested in olfaction.

ACKNOWLEDGMENTS

It is a pleasure to acknowledge the help and stimulation I have received. It was Carl Pfaffmann who first encouraged me to channel my interest in psychophysics into the study of taste and smell. Gösta Ekman supported my tendency to focus this work at "the psychological level," as he described it, at a time when that was far less popular than it is now.

The data described in the book are to a large extent the research of students and younger colleagues such as Birgitta Berglund, Ulf Berglund, William S. Cain, Harry T. Lawless, Thomas Lindvall, and Robert G. Mair, all of them undoubtedly to be heard from in future developments of these topics.

The preparation of the manuscript could not have been completed without the help of Ann Marie Clarkson in preparing drawings, Elizabeth A. Engen for criticisms of ideas and writing, Caroline M. Healie in tracking down and obtaining references, and Helen J. Shuman for typing the manuscript and keeping it organized.

chapter 1
INTRODUCTION

Before odor pollution became a widespread problem, the sense of smell received only cursory attention in scientific treatments of its psychology and physiology. By contrast, writers like C. S. Lewis held this modality in higher esteem. As Lewis commented in a discussion of the value of perceptual experience: "Of landscapes, as of people, one becomes more tolerant after one's twentieth year. . . . We learn to look at them not in the flat but in depth as things to be burrowed into. It is not merely a question of lines and colors but of smells, sounds and tastes as well: I often wonder if professional artists don't lose by seeing it in eye sensation only [Lewis & Executors of C. S. Lewis, 1966, 234]."

A person who has lost the sense of smell (an *anosmic*) is deprived of more than pleasurable sensations, for it has been argued that as the environment becomes more polluted what the nose knows may have to be taken seriously (Moncrieff, 1966). One reason that the sense of smell has been given less consideration than other modalities, especially sight and hearing, is implicit in its characterization as the "subtle sense" (Bienfang, 1946). When contemplating the roles of one's senses, one is apt to think about odor in terms of detecting skunks, and hearing in terms of listening to speech, or, generally, the experience of conscious events. By definition, one is not aware of nonconscious sensory inputs, which nevertheless may be important in daily life. However, when odor experience is missing, as when one is suffering from a bad cold, and foods turn into lukewarm mush varying only in a few gustatory (taste) qualities, the contribution of smell to taste is appreciated. It has been suggested that the consequences of permanent anosmia may affect not only one's appetite but one's love life as well.

Cognitive psychology deals exclusively with auditory and visual perception, even in discussing memory. Related to this is the emphasis in contemporary cognitive psychology on human voluntary behavior and experiments that emphasize words, numbers, and letters as stimuli. But even the knowledge of the rules of one's native language is unconscious (Chomsky, 1957). Stimuli to which people might respond unconsciously or nonverbally tend to be ignored, and such an orientation suggests the conclusion that perception of odor is not important to people. Freud, certainly not a cognitive psychologist, believed that some of the most important facets of life do not reach consciousness, and Wiener (1966, 1967a, b, 1968a, b) makes a special case for odors or chemical messengers as potentially important both to the individual and to interpersonal social relationships. That odor perception is not important to people is a relatively recent idea, loosely defined and not based on any kind of objective analysis of behavior. Novelists and other observers of human behavior often describe the important role of odor perception in human experience, and it is now getting more and more attention by scientists.

The emphasis in this book will be on human perception, but contributions and hypotheses based on animal observations will be discussed, because it is inherently interesting to compare humans and animals and also because animal research in the neurophysiology of the olfactory system is indispensable. No attempt will or indeed could be made to restrict the discussion of smell to what is mediated by the first cranial or olfactory nerve. Physiologists do this in order to distinguish between stimulation mediated by this nerve and that mediated by the trigeminal nerve (which involves experiences of pain and irritation and not olfaction in the physiological sense). Since the strength of an odor may determine the extent to which different nerves are stimulated (Cain, 1974a), human odor perception cannot be limited on that basis, except in very special cases. Human introspection or other behavior of an intact organism does not provide a valid index for such a neurological distinction. *Odor* or *smell* refers to odor experience; that is, sensory experience or other behavior presumed to be stimulated by the olfactory system (though the trigeminal system may also be involved, as when one is exposed to ammonia). *Odorants* are the distal olfactory stimuli from which molecules emanate. That is, the sensation of odor is the result of stimulation by an odorant. (In nonpsychological literature, including some instances in this book, *odor* refers to both the proximal resulting response and the distal source.) The discussion of the interrelatedness of the senses (Chapter 9) will have more on this topic, but throughout this book a broad psychological definition of the sense of smell is implied. Along this line, a broader treatment is implied by the use of the word *perception* in the title rather than *sensation*, which is usually restricted in psychology to the experience of simple attributes; perception is of whole objects.

The discussion will not stop at this, but will look at odor experience in a broader perspective. Functionally, smell may be to emotion what sight or hearing is to cognition, and one could justify classifying the sense of smell under the more general rubric of motivation rather than information processing. Such thinking is part of a renewed interest in this modality. For example, the significance of body odors in interactions between animals is generally recognized, and it has been suggested that body odors are also important in human interaction. That unnoticed odor signals may be used to repel or attract others, or even influence another person's hormone balance, has been argued, but not demonstrated. Perfumers advertise odor as "the secret weapon of sex." So the approach in this book will be broad, biased toward human application; in brief, it will cover what odor perception does for and to a person.

The study of the sense of smell, like that of experimental psychology, has a long past but a short history (Boring, 1942). Any human characteristic may be psychologically relevant and mention of it may be found in earliest recorded history. McCartney (1968) suggests that Theophrastus' work "Concerning Odours," written in the second or third century B.C., may be the first of its kind. This scholar also wrote about rocks and plants and treated all of these things at a very general level, judged by present standards. Most of the old literature is anecdotal and descriptive and of little explanatory value, and I shall not dwell upon it.

The remark that the study has only a short history refers to the fact that the scientific study of the perception of odor is only about 100 years old. Boring (1942) writes that it was the erudite Dutch physiologist Zwaardemaker who created interest in the psychology of smell with his book *Die Physiologie des Geruchs* in 1895. One clear sign of the book's importance is that it was revised and published in another language, French, under the title *L'Odorat, in 1925.* However, many questions had been raised before the publication of Zwaardemaker's book, and in that sense the study of smell has had a long past. These questions are still interesting today, and will be discussed in more detail, both in this chapter and in later chapters along with contemporary questions.

Odor Acuity

Most discussions of olfaction are likely to state that, compared with vision and audition, the sense of smell is dull. This comparison is usually accompanied by another in which the sense of smell of humans is said to be poorer than that of animals, often the dog, which is presumed to need and possess a keen sense of smell. Next, it is usually suggested that the dullness of human odor sensitivity indicates its lack of significance for humans, as though it was in fact a sign of having become civilized; also suggested is that the present poor olfactory acuity of humans is the result of a steady decline, which started when our animal ancestors came out of the water, then became upright, eventually going up in the trees. Dröscher (1969) presents a very readable popular account of the marvels of odor perception in animals. Presumably, odors tend to be concentrated near the ground and therefore human beings are preoccupied with vision and audition. Olfaction is more involved in visceral and emotional activities than in sensory information transmission.

The potential importance of the sense of smell is being given new consideration today, because of evidence that chemical messengers may play an important, perhaps decisive, role in the social and sexual behavior of human beings as well as animals. Not all the evidence supports the view just noted that the human sense of smell is poor. In some cases the human sense of smell is very keen. For example, in some cases it may be unsurpassed as a chemical detector (Wright, 1966). An average person can detect the rotten egg odor of hydrogen sulfide, the garlic odor of thiocresol, or the skunk odor of ethyl mercaptan if only 10, 1, or .5 ml, respectively, of the substances are dispersed in 10,000 liters of air. An example more easily grasped intuitively is that one may be able to detect one small drop of perfume diffused throughout a whole house. Of course, in one drop of perfume there would be an enormous number of molecules, so that one does not have a good frame of reference against which to judge such data and to compare them with acuity data in other modalities involving other physical dimensions. Nevertheless, an average person is good at detecting odors when compared with so-called physical sensors such as a smoke detector.

There are some "notable noses" (Bedicheck, 1960, p. 53), people who are assumed to be exceptionally sensitive to odors. Among them would be Helen Keller, wine tasters, perfumers, and fish-smellers and other assorted organoleptic analysts. The special experience and condition of these people make it sound quite plausible that they are very sensitive, but the information tends to be based on controversial, limited, and nonobjective evidence. Olfaction generally gets little attention and odor perception may be so often repressed that when it has been shown to play a significant part in a certain situation or for a particular person, its sensitivity may be exaggerated or the owner of the particular nose may be suspected of having special olfactory power.

One problem with notable noses, already alluded to, is that the information about them is always anecdotal. There are no measures of olfactory sensitivity similar to an audiometer, for example, in the case of hearing. This is a problem of applied psychophysics, where *psycho* refers to sensory or experienced subjective magnitude—for example, loudness or odor intensity—and *physical* refers to sounds, pressures, or concentration. It was Fechner (1860/1966) who originated psychophysical methods for the purpose of describing the relation between internal experience and external physical stimulation, between body and mind. Fechner's psychophysics has since had an important impact on both psychology and physiology, because it provided objective methods for measuring sensation, which previously had been considered private and unmeasurable. In particular, it has had an impact on applications to practical measurements of acuity and sensitivity in various modalities—for example, in measuring hearing loss. The technology needed for similar evaluation in the case of smell is known. As we shall see, the development of psychophysics is one of the key factors in the advancement made in the study of smell. Unfortunately, it has not yet affected those involved in practical application, and measurement of odor sensitivity and acuity is still made very crudely. Thus, one reads anecdotes about the sensitivity of special people such as Helen Keller, but one never sees any results from objective tests, because there are none.

The notable nose is probably bred rather than born; we could all improve our ability to discriminate odors if we trained our noses, although there is no obvious need to do so. It is rumored that the business of experienced noses became an issue in the Vietnam war. The Vietnamese supposedly used their sense of smell to detect the whereabouts of machinery and other things. On the U.S. side, experts in olfaction were called together in an olfactory brainstorming session, which resulted in the development of an artificial nose (analogous to a smoke detector) that could detect the body odors of Vietnamese guerrillas.

To use an example closer to home regarding the value of a trained nose, an article on the first page of the *Wall Street Journal* (October 7, 1975) entitled "The Sweet Success of Smell: Mr. Weber Sniffs Way to Top" deals with a Food and Drug Administration expert who has specialized in the detection of rotten fish. He is a trained chemist who uses his sense of smell rather than chemistry because there is no acceptable test for rotten fish, and chances are it will take some time to develop one that can undercut the human nose in either sensitivity or efficiency. When the article was written, Mr. Weber was still working at 70, after 32 years of experience. (He did have apprentices.) This human if not humanistic approach is also economical of time and is reputed to accomplish in 2 hours time what would require 2 days for comparable chemical testing (see Hopkins, 1975).

What Is the Odor Stimulus?

The relatively poor performance of chemical tests bring us to what might be the most crucial question of all. That humans experience odor is a fact, but what causes this experience remains a puzzle. Is it chemical or physical? The answer is not clear. Discovering what part or attribute of the molecule of an odoriferous substance causes a change in the olfactory receptor, which in turn leads to the experience of odor, would be worthy of a Nobel Prize. We know that the smell of rotten fish is associated with a chemical change produced by decomposition, but what in particular the chemicals involved do is not known. Some very similar chemical substances have quite different odors—for example, *d*-carvone smells like caraway and *l*-carvone smells like spearmint. Both are stereoisomers; their formulas are the same but the two molecules are mirror images. This great qualitative change in sensation from a small change in the stimulus is reminiscent of categorical perception assumed to be indicative of distinctive feature detectors in, for example, speech perception, where discrimination is relatively poor within a sound class like the syllable "pa" compared to discrimination at the boundary between sound classes like "pa" and "ta" (Studdert-Kennedy, Liberman, Harris, & Cooper, 1970). In olfaction it is not possible to produce the small, graded dimensional changes required for an experimental test. On the other hand, some substances with quite different formulas smell alike—for instance, exaltone and musk ambrette have quite different molecular configurations, but they both have a musky odor. The problem has defied any simple chemical or physical solution. Table 1.1 gives a classification of the various ideas proposed to explain what stimulates the olfactory receptor and what accounts for the experience of different odor qualities.

The idea that the stimuli could be particles or atoms dates back to the Epicureans (Cain, 1978a). Two thousand years later this theory was supported experimentally; more precise specification of the number of molecules in gaseous odor samples became possible. Quantitative psychophysical studies of odor samples got under way, but the stimulus problem has not been solved yet. During one phase of this development, it was assumed that no scientific work could be done without this knowledge. However, as S. S. Stevens (1951) wrote in the introductory chapter to the *Handbook of Experimental Psychology*, the basic problem of psychologists is to work toward the definition of the stimulus. Thus, the present problem was channeled into a concern for doing testing under the most careful laboratory conditions so that the significant factors could be determined. In audition this led to the anechoic chamber (that is, a chamber free from echoes); in vision, the darkroom; and in olfaction, the olfactorium (camera

Odor Classification

Table 1.1
Types of Theories Explaining the Olfactory Stimulus[a]

Class	Rationale
Radiation and vibration theories	Molecules vibrate at different rates.
Mechanical theories	Force is associated with the weight and momentum of the odorant molecule.
Stimulus pattern theories	Different molecules stimulate different regions of the olfactory epithelium.
Chemical theories	Some chemical reaction occurs between the molecule and the receptor.
Steric theories	The shape of the atom configuration interacts with the receptor.
Phase boundary theories	Interface occurs between the molecule and the receptor in solution.
Enzyme theories	Molecules affect enzymes within the receptor.

[a] After F. N. Jones and M. H. Jones (1953).

inodorata), a smellproof or odorless room. One of the earliest efforts of this kind was a chamber established at Cornell University with the support of Joseph Seagram and Sons (Foster, Scofield, & Dallenbach, 1950). This was a well-ventilated but airtight glass chamber in which controlled amounts of odor were presented to a subject. The chamber was large enough to hold observers or subjects and other apparatus and included provisions for deodorizing both the equipment and the experimental subjects with showers, sterilizers, and the like. As we shall see (Chapter 3) present techniques do not control the subject's environment that severely.

Odor Classification

In no other modality has classification dominated research as it has in the case of the sense of smell. The usefulness of classification in color vision has been another factor motivating odor classification. The hope is still that after first having ordered all odorants (or as many as one can manage to sniff) into as few groups or classes as possible based on the similarity of their odors, one would then be able to specify the physical or chemical similarity of the substances in each class. (See Table 1.2 and Figure 1.1.) That similarity, in turn, would provide the clue to the nature of the stimulus, and then finally one would be in a position to study the nature of the neurophysiological transduction mechanism. Knowledge of this phenomenon would, of course, represent a fundamental answer to

Table 1.2
Linnaeus' 1756 Odor Classification

Class[a]	Example
I. *Aromaticos*	Carnation
II. *Fragrantes*	Lily
III. *Ambrosiacos*	Musk
IV. *Alliaceos*	Garlic
V. *Hircinos*	Goat
VI. *Tetros*	Certain bugs
VII. *Nauseos*	Putrefying flesh

[a] The extremes (I, II, VI, VII) represent pleasant and unpleasant odors and the middle categories (III, IV, V) represent more neutral odors. Linnaeus used flowers as examples of each class. (See Harper, Bate-Smith, & Land, 1968.)

what the nose knows. One cannot ignore this methodological approach, but be forewarned that although there has been some progress in developing a common glossary (Harper, Bate-Smith, & Land, 1968), no agreement has been reached on the question of what odors are similar or how many classes there might be among the estimated 400,000 odorous substances (Hamauzu, 1969).

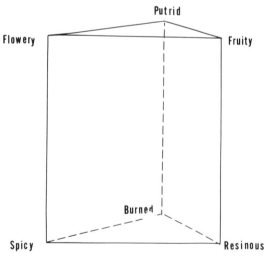

Figure 1.1 Henning's odor prism. A model believed to depict the psychological ordering of different qualities located, not in the prism, but on its surfaces. (After Henning, 1924.)

Odor Mixture

The study of odor perception has no truly great problem solver comparable to Helmholtz in vision or Bekesy in hearing. On the physiological side there was Zwaardemaker with *Die Physiologie des Geruchs* published in 1895 and *L'Odorat* in 1925, and on the psychological side there was Henning with his opus *Der Geruch,* published in 1916 and revised in 1924. Henning's classification of odors in the geometric form of a prism (Figure 1.1) is still the most widely quoted model of the "odor space." Henning and Zwaardemaker apparently did not get along well, and their main bone of contention was the problem of compensation. In addition to inventing an olfactometer for studying the olfactory stimulus and other psychophysical matters, Zwaardemaker was the first to suggest that two odors, each distinct, would compensate for each other, making a mixture of the two weaker rather than stronger. He also showed that compensation occurs when odorants are presented singly, each in its own vessel, and inhaled through separate nostrils in a so-called dichorhinic mixture. In the extreme case, according to Zwaardemaker, complete compensation or neutralization takes place, with odor disappearing altogether—a possible dream come true for the deodorizer business, although neutralization was likely only if the components were not too strong.

Henning also worked on odor mixtures in connection with his odor prism (Figure 1.1). He was interested, for example, in determining the perceptual result of mixing two odors, each typical of its class; he concluded that the resulting odor tends to be a unique percept blend in which both components can be smelled. Henning never found any evidence of compensation, let alone neutralization, and believed that Zwaardemaker's findings could have been the result of a chemical rather than a physiological effect.

Although complete compensation or neutralization has not been established as a fact for any combination of odors, Henning's conclusion was too general. It is a well-established experimental finding that the perceived strength of a mixture of odors is generally less than that predicted from the sum of the individual intensities of the components. The methodology and results of such scaling will be discussed in more detail in Chapter 3, for this is a very active field of research today. Understanding odor mixtures is necessary in the study of deodorizing and air pollution, and significant for odor theory in general. For example, a startling and controversial claim has been made by researchers at Monsanto, who say that they can, with a patented "fresh air" smell, counteract a wide variety of environmental malodors without affecting pleasant odors. Zwaardemaker might not be surprised at such findings, but this contemporary approach to odor neutralization also has its critics (see Pantaleoni, 1976).

Odor Coding

What has been discussed so far bears on the next question—an important one—of how the odorant initiates a change in the olfactory receptors. Boring (1942) writes, "Zwaardemaker spent a large part of his professional life looking for new physical principles that would give him a key to smell by an understanding of the true nature of its stimulus [p. 438]." He did not succeed, and neither has anyone else; but improvements in recording techniques, microscopes, and psychophysics now permit rapid advances in the field. Anecdotes from physicians and writers—see, for example, Ellis' (1928) popular work—used to predominate in works on the sense of smell. The lack of stimulus control kept serious experimental psychologists away. Even after World War II, it was considered a sign of either courage or lack of grantsmanship to devote oneself to the study of odors. Boring (1942, p. 437) estimated that in the 1940s the understanding of smell had reached the level of understanding sight and hearing had in 1750.

It has been argued that most of what is known in science has been acquired since Sputnik. The gap estimated by Boring has surely decreased, but it is still there. Writers discussing theories of olfaction still feel obliged to mention every theory ever dreamed of, essentially because there was not enough information to reject anything sounding plausible. As indicated in Table 1.1, F. N. Jones and M. H. Jones (1953) found so many theories that they had to group them into seven classes, which they evaluated, they then concluded by saying, "It is apparent that more speculation than fact has been abroad in the field, and that there is little point in pursuing theory-spinning further without resort to adequate testing of hypotheses [p. 235]."

The number of theoretical contenders now may be counted on one hand. Most contemporary theories suggest that odorant molecules come in contact with the receptor cell, but they differ on how this takes place and the nature of the interaction. The field has come a long way from the Epicureans' atom theory, but the nature of the transduction is not yet understood. Progress has been made in this field partly because of technological advances, but as Cain (1978a) points out, also because neurophysiologists finally got into the act, beginning in the late thirties. This aspect of the problem will be pursued in the next chapter, but it is necessary first to ascertain the psychological nature of the response to an odorant that has affected receptors and passed the threshold of awareness.

Hedonics

The most important aspect of an odor has generally been believed to be its hedonic effect. To some, odor is practically synonymous with pleasure or, rather, displeasure (Figure 1.2). It is generally assumed that foods smell pleasant because they are wholesome and that harmful substances, in-

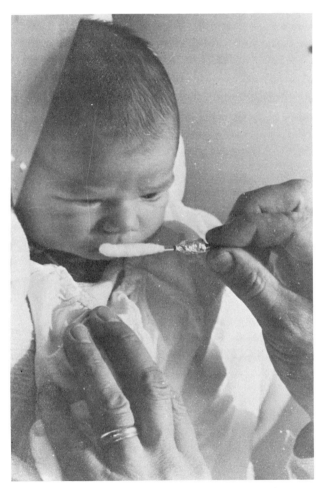

Figure 1.2. It is debatable whether or not newborn babies discriminate between pleasant and unpleasant odors and use the sense of smell to select food; they seem suspicious of any odor regardless of its pleasantness to adults. (A photo by Mort Blender from the Neonatal Behavior Laboratory at the Women and Infants Hospital of Rhode Island.)

cluding food that has gone bad, will have a disagreeable smell and will therefore be rejected. Under such circumstances the olfactory system is thought to play its most important adaptive role in responding in accordance with the "wisdom of the body." The degree of pleasantness of food odors will vary with a person's state of satiety and hunger.

The historian Kern (1974) writes that there was a particular interest in this hedonic problem early on,

> because of the popular theory that odors emanating from putrefying organic bodies caused disease—in particular cholera, which had erupted in epidemics in 1831 and again in 1848. The desire to eliminate miasmatic origins of disease became linked with the campaign to provide adequate ventilation in hospitals, factories, schools, and homes. The "Great Stench" that choked London in the summer of 1858 gave support to the efforts that Florence Nightingale had made around that time to improve ventilation in hospitals. The English epidemiologist, William Budd, commented on the historical significance of the Great Stench, "never before had a stink risen to the height of an historic event" [pp. 816–817].

Interest in this topic has been renewed. Actually no evidence has been revealed that merely smelling an odor will produce disease. But everybody has experienced revulsion and the need to escape malodorous situations, and the definition of health has been expanded to include purely psychological in addition to physical well-being. Whether for real or imagined reasons such experiences seem to be increasing. On the other hand, during the student upheavals of the late 1960s and early 1970s, there were some who gave up deodorants, claiming that body odors are natural and that deodorants are therefore unnatural constraints. One might have been inclined to believe that this occurrence was a first, but references to this attitude go back at least 100 years, and it was, according to Kern (1974, p. 821), then associated with *Freikörperkultur* and nudism.

At about the turn of the century, there were others who warned that civilization tended to eliminate odors and that this would have a negative effect on human sexuality and, in general, aesthetic life. Somewhat later, but clearly for the same reason, Freud (1962) stated that this kind of repression could cause mental illness. The most detailed analysis of the medical aspect of odor perception has been made by another scholar and sexologist, Havelock Ellis (1928). Among other things, Ellis proposed that, although strong body odors are normally thought to cause repulsion, the response will be modified in terms of the relationship between the perceiver and the perceived. If the latter is a lover, his or her body odor may be acceptable, whereas the same degree and type of odor from a stranger would be repulsive. Subtle odors are generally preferred, nevertheless. Plautus, as quoted by Bedicheck (1960, p. 125), wrote about such smells that, "Then smells a woman purely well, when she of nothing else doth smell."

Hedonics

There are also frequent references in the old literature to racial differences in body odor—for example, Japanese finding the odor of Westerners unpleasant (Ellis, 1928). These differences were presumed to be unrelated to diet and hygiene, which are obviously involved in body odors. For example, the odor of breast-fed human babies is reported to be very different from and more pleasant than their odor when they are given formula. In general, new evidence of individual idiosyncratic body odors analogous to fingerprints adds to the intrigue. Each of the three males in our house has his own unique smell, according to its sole female. It has been demonstrated that dogs can tell people apart by their odors, even in the case of identical twins, whose odors are apparently more similar than those of others (Kalmus, 1955). However, a dog trained to detect the odor of one twin can track the other twin, because their odors are also more similar than those of a nonrelated pair. A related and fascinating possibility is that the olfactory and immunological systems may be linked. Congenic mice can discriminate one another's histocompatibility complex on what appears to be an olfactory basis, independent of sex differences and circumstances of mating (Yamazaki, Yamaguchi, Baranoski, Bard, Boyse, & Thomas, 1979). According to Thomas (1974),

> it has recently been learned that the genes for marking of self by cellular antigens and those for making immunologic responses by antibody formation are closely linked. It is possible that the invention of antibodies evolved from the earlier sensing mechanisms needed for symbiosis, perhaps designed, in part, to keep the latter from getting out of hand [pp. 40–41].

Are there inherently pleasant and repulsive odors? It seems clear that displeasure dominates in odor experience. Only one-fifth of the nearly half-million odorous compounds we know of are judged pleasant (Hamauzu, 1969). The reason for this is partly that the function of the sense of smell is to alert and warn persons or animals, to put them in a state of arousal. Steiner (1977) has shown evidence of inherent preference for tastes from the facial expressions made by newborn human infants. For example, to the sweet taste of sucrose they seemed to look pleased but grimaced to the sour and bitter tastes of citric acid and guinine sulfate. However, similar tests with odors intended to elicit responses of pleasure and displeasure were not as clear cut. But familiarity will undoubtedly change such hedonic responses for those items that actually do bring pleasure (Engen, 1974). At birth a child shows little evidence of liking odors judged pleasant by adults; the ability to detect odors is present at birth but the hedonic scale is largely, perhaps totally, a developmental phenomenon, possibly the result of physiological maturation but probably the consequence of experience. It is very difficult to convince oneself of that, and parents are

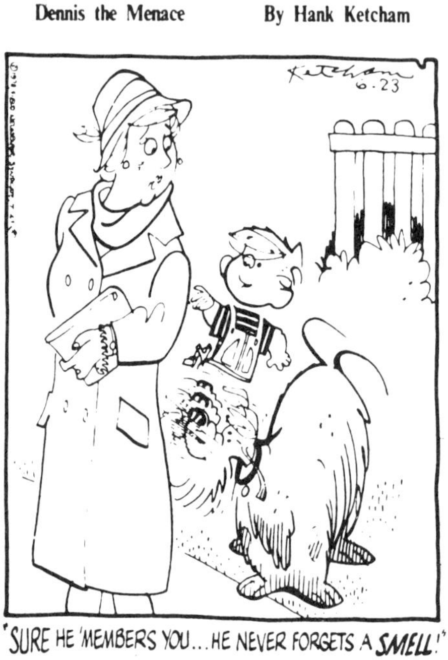
Dennis the Menace
By Hank Ketcham
Ketcham 6-23
"SURE HE 'MEMBERS YOU... HE NEVER FORGETS A SMELL!"

Figure 1.3. Dennis the Menace® cartoon used by permission of Hank Ketcham.

sometimes astonished at how tolerant their 1- and 2-year-old children are to odors such as dimethyl disulfide, which the parents find repulsive (L. P. Lipsitt *et al.*, 1975).

Odor Memory

Naming odors is a difficult cognitive task. The initial experience in smelling an odor may be hedonic, a feeling state rather than a sensation, whereas in vision the sequence may be the reverse (Achiles, 1929). The stress on feeling is thought to be the reason that odor memory seems to be exceptionally good. For example, it is customary to quote Proust (and his memories associated with *petites madeleines*) and other artists who have claimed that odors are not forgotten to the same extent as other sensory experiences. Experimental evidence supports these observations in the case of the permanence and persistence of the phenomenon of "bait shyness," the rejection of food that was previously poisoned and made the animal sick. Merely overindulging in food or drink may have the same effect on humans. Subsequent pleasant experiences with such foods do not readily change these dislikes (Lawless & Engen, 1977a), and food aversions may last a long time; they may become permanent. Although taste is primarily involved, odor perception plays an important role in the recognition of food.

There is now a good deal of laboratory evidence supporting the Proustian hypothesis for recognition of odors, but that is only one test of memory. It is not clear whether or not one can actually recall odors. Proust described the *recognition* of odor—that is, knowing that an odor being experienced is one that was experienced on an earlier occasion. But recall is a different test, requiring the person to bring back odor sensations from memory storage without any external aids. Industry is taking advantage of the ability of people to recognize odors with the development of "microfragrances," where odorous molecules are trapped in a strip of paper. When the paper is scratched the bubble is ruptured and the molecules released.

But how can one conjure up an odor without any kind of prompting? Some people claim they can do this—for example, that just thinking about a lemon will bring back its odor, but others do not think they can do this. When I try, what comes back is the visual image of the lemon or other associations to lemon, but not odor experience. This brings us to the brink of what contemporary research can accomplish, for there seems to be no way to test this problem objectively, and for the most part this book will be oriented objectively.

Organization of This Book

We begin with a broad outline of the anatomy and physiology of the olfactory system (Chapter 2). The bulk of the data here is of course based on animal and insect research, but comparisons to humans can certainly be made. The basic question concerns what is known about the neural correlates of psychological attributes of odor intensity, quality, and hedonic experience. This topic goes hand in hand with psychophysics (Chapter 3), the quantitative study of the relationship between physical stimulation and conscious experience and its representation in the nervous system. The central problems of psychophysics are (a) odor sensitivity as in detection or threshold experiments; (b) odor quality, as in classification experiments; and (c) the way in which perceived odor intensity or strength grows as a function of increased odorant concentration.

The chemical senses, as they are traditionally called, and in particular the sense of smell, are believed to be characterized by adaptation (Chapter 4) when exposed to a constant suprathreshold stimulus. Although the odor of a certain room may be quite distinct as one enters it, the odor will rapidly wane, according to traditional beliefs. The classic literature (for example, Zwaardemaker, 1895; see also Pfaffmann, 1951), states that complete adaptation and diminution of odor should require no more than about 5 minutes. We now know that this conclusion must be revised, partly because new and better psychophysical methods have provided better data. Also, the problem of adaptation has been studied from a broader perspective and includes not only adaptation and cross-adaptation, the effect of one odorant upon another, but also facilitation of one odorant on the detectability of another and alliesthesia, the effect of one's internal physiological state on the perception of intensity versus pleasantness of odors.

There are people whose sensitivity to odors is very poor and sometimes totally absent, a condition called *anosmia* that may be caused by accidental damage to the olfactory system or by genetic factors (Chapter 5). Although given scant attention by a few physicians (e.g., Schneider, 1974), this condition is obviously of medical significance, and interest in it is on the rise. At one extreme is the observation on the excellence of odor memory (Chapter 6), and at the other extreme are medical reports of anosmia and Freudian repression of odor. A related cognitive problem is whether or not the sense of smell is analytic, like hearing, by which we can pick out the various instruments of the orchestra, or synthetic, like vision, in which mixtures of colors produce sensory results different from any of their components. Knowledge of this is of course of special interest to perfumers and deodorizers, and these problems will be discussed under the general subject of odor mixtures (Chapter 7).

Why are some odors liked and some disliked? Individual and cultural differences and development of such hedonic preferences in a person or society are considered by many the most interesting aspect of the odor problem (Chapter 8). The suggestion that there might be human pheromones, or airborne hormones that function as aphrodisiacs, is especially exciting and perhaps the primary reason for the upswing of interest in olfaction. A more mundane consideration is the effect of odor on taste or flavor and the effect of other senses on smell (Chapter 9). Is the popular notion valid that when one is temporarily functionally anosmic, because the nose is stopped up from a bad cold, that one also suffers "taste-blindness" (hypogeusia)?

Finally, I shall point to some unsolved problems but also to potential practical applications of odor research—such things as making artificial noses and preventing pregnancies (Chapter 10). In the last chapter I shall bring together the main principles of the function of the human sense of smell (Chapter 11).

chapter

2

GROSS ANATOMY AND PHYSIOLOGY

Odor is experienced when molecules of odorants reach the olfactory epithelium at the top of the nasal passages or nares (Figure 2.1). It is in this membrane that the olfactory receptors are located. I shall take up each of the anatomical structures and physiological mechanisms in turn, with the purpose of providing sufficient introduction to the mechanisms and problems referred to in other chapters.

The Nasal Passages

The nares are separated by the septum, which provides two separate channels of information for the observer. This has stimulated research on the lateralization of odor sensitivity, but results have been confusing (see Koelega, 1979; Youngentob, Kurtz, Leopold, Mozell, & Horning, in press). Some have found the left nostril to be more sensitive than the right, but Koelega (1979) suggests that the results might have been confounded because emotions associated with odor perception may be better represented in the right hemisphere.

The nasal cavity as a whole consists of three parts; an anterior naris, a central chamber containing the epithelium and the receptors, and a posterior naris through which air inhaled through the anterior naris exits into the nasopharynx behind the soft palate. Odorants may reach the receptors through the latter route, for example when one is eating. Humans can, of course, breathe through the mouth, and the nares may be considered

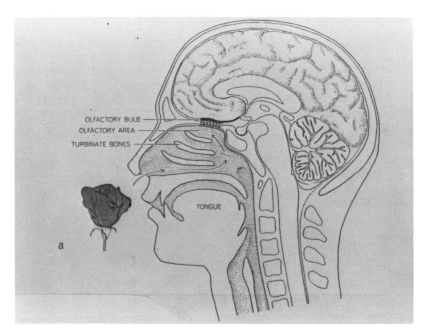

Figure 2.1. Gross anatomy of human olfaction. The right nasal passage is shown, including the turbinate bones, the epithelium or olfactory area, and the olfactory bulb. (From "The Stereochemical Theory of Odor," by J. E. Amoore, J. W. Johnston, Jr., and M. Rubin. Copyright © 1964 by Scientific American, Inc. All rights reserved.)

specialized for olfaction as well for cleaning and warming the air. However, the structural design of the nasal cavity as a whole is complicated, and in each cavity there are folds or conchae keeping air from flowing smoothly through the nostril. The number of conchae varies according to species, from zero to more than 10 (Parker, 1922). In general, this turbinate system seems to become more complex as one moves along the phylogenetic scale from amphibians to mammals. Yet among primates there is a decrease in complexity from prosimians to anthropoid apes (Loo, 1973). Is this feature correlated with the importance of the sense of smell for animals?

It is difficult to predict how many molecules inhaled from the air outside the nose actually reach the olfactory slit or cleft where the olfactory epithelium is located, and not surprising to find that the amount of vapor reaching the epithelium is not a simple linear function of the concentration and flow rate presented (Aharonson, Menkes, Gurntner, Swift, & Proctor, 1974; Tucker, 1963). The epithelium lines the walls of the slit and covers 2.5 cm^2 or more on each side in humans. Statistical estimates of the number of molecules of various odorants getting to the slit through normal breathing (about 250 cm³/sec per nostril) have been made (De Vries & Stuiver, 1961). At that rate, it is estimated that 5–10% of the air inhaled will pass through the olfactory cleft; for higher flow rates, the figure is no more than 20%. Flow rate is an important factor in determining the strength of odors. When the flow rate is manipulated by an experimenter using an olfactometer to present an odor, the perceived odor intensity increases as the flow rate increases, with concentration held constant (Rehn, 1978, 1979). However, Teghtsoonian, Teghtsoonian, Berglund, and Berglund (1978, 1979) found that a flow rate that varied as a result of the vigor of the subject's sniff did not influence perceived odor intensity or, presumably, the number of molecules available at the epithelium. The authors suggest that this result must be another instance of perceptual constancy (discussed more fully in Chapter 10).

Before the odorant gets to the epithelium, molecules may be lost on the way because of adsorption on the other surfaces in the nasal cavity. Of the molecules reaching the cleft, only a fraction will be adsorbed into the mucus, and not all of the molecules in that fraction will diffuse through the mucus. The rate of diffusion varies for different odorants and diluents, and only some of them will reach and interact with the receptors. Desorption takes place after this interaction, when the molecule presumably will be moved out past the nasopharynx by the air current in the nostril (De Vries & Stuiver, 1961; Moulton, 1976b).

The Epithelium

The epithelium is a tissue about 75 μm thick, which covers the olfactory cleft. Embedded in this tissue are three kinds of cells: receptor cells, supporting (or sustentacular) cells, and basal cells. The tissue is covered by mucus resulting from secretions of the supporting cells and Bowman's glands that penetrate the epithelium and subepithelium connective tissues. The exact nature and function of these secretions are not yet known (Moulton & Beidler, 1967). The mucus is a viscous, colloidal substance about 20–35 μm thick in the frog. It contains a number of poorly understood particles (Reese, 1965). The mucus-covered epithelium membrane is called the olfactory mucosa and differs from other mucus-covered membranes in the nasal cavity in that it contains the olfactory receptors and has a characteristic yellowish brown color due to a pigment. It is not easily observed because of its location, but it is known to be a good medium for adsorption and transmission of molecules (Beets, 1970). Supposedly, the molecules may remain in the mucus after they have affected the receptors, but their effect at this stage is not clear. (See Figure 2.2.)

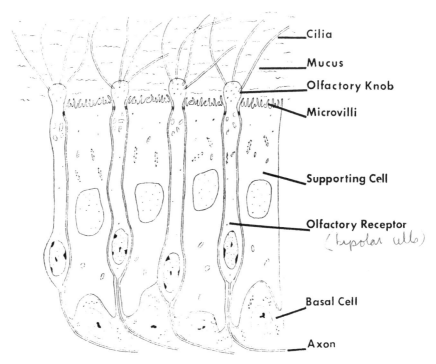

Figure 2.2. The structures of the olfactory epithelium. The trigeminal nerve endings located between the supporting cells and below the surface are not shown. (From Moulton & Beidler, 1967.)

The Olfactory Receptors

The olfactory receptors as observed in the electron microscope are not morphologically distinct types, as are, for example, rods and cones in the retina. The olfactory receptors are bipolar cells in which the dendrite and axon connect directly to the cell body, which lies between supporting cells in the middle of the epithelium. The dendrite is only about 1–2 μm in diameter and extends to the epithelial surface, where it terminates in a swelling called the *olfactory knob*. A rough estimate is that humans have about 10 million receptor cells altogether in an area covering as much as over 5 cm², which is more than for any other sense modality except vision. There is a rapid turnover of these cells; they are continuously being renewed, and in just a few days will mature and degenerate (Graziadei & Graziadei, 1978). More receptors and larger areas covered by the olfactory epithelium are found in animals assumed to have greater sensitivity; dogs for example, apparently have over 200 million receptors and have been demonstrated to be more sensitive than humans (Marshall & Moulton, 1981). The density of receptors varies from species to species and the boundaries of the olfactory mucosa are not distinct (Moulton & Beidler, 1967; see also Laing, 1975).

On each knob there are several cilia up to 200 μm long and .2 μm in diameter. They provide a surface for interactions with molecules. It has long been believed (Parker, 1922) that these olfactory cilia contain the olfactory receptor sites where this interaction takes place, and this remains a viable hypothesis. The olfactory cilia move in different directions and more slowly than neighboring respiratory cilia, which move in a more rapid and synchronized manner. Some researchers believe that this motion may be involved in the displacement of the odorant molecule from the receptor site (Bronshstein & Minor, 1973).

Outside the olfactory area the mucous membrane is columnar epithelium. It is important to keep in mind that olfactory receptors are not alone in the nasal passages (Tucker, 1971). Of particular interest are the free nerve endings of the common chemical sense that responds to irritating chemical substances. Fibers from these receptors join fibers from other areas of the face and head in the trigeminal or fifth cranial nerve. These receptors are stimulated by many substances, such as automobile exhaust (Degobert, 1979a), which may also stimulate the olfactory receptors, and sensations from the trigeminal and olfactory receptors cannot always or easily be distinguished by a human observer. However, the trigeminal system is not well understood (M. H. Jones, 1954). An interesting study pertaining to this has been done on humans having unilateral destruction of the trigeminal nerve, leaving them one normal nostril and one without trigeminal input. Psychophysical data obtained from them show clearly that what is judged to be odor intensity is partly the result of trigeminal stimulation (Cain, 1974a). The olfactory and trigeminal systems also mutually inhibit each other, such that an odorant may mask an irritant and vice versa (Cain & Murphy, 1980). This is a case where work with animals is helpful in reaching a clearer picture of the underlying physiology. For example, much of this work has been done on the frog, whose olfactory epithelium is similar to that of humans and relatively easily exposed surgically.

Also of interest is the vomeronasal organ, a tubular structure located in the septum (Wysocki, 1979). This organ is evidently also called Jacobson's organ, after a Danish physician who was among the first to study this region. However, it is not entirely clear whether the two terms actually refer to the same organ. Much of what is known about it is based on circumstantial evidence, and its form varies in different animals. The vomeronasal system is separate from the olfactory system; it has its own connection with the central nervous system, and its cells are functionally similar to olfactory cells. This system may serve the specialized function of mediating responses to pheromones, but it is not clear whether or not this inconspicuous organ exists for humans and other higher primates. It is evident in snakes and lizards, and may be rudimentary in the young human child but seems to be absent in the human adult. This is a specially interesting problem considering the related question of whether human pheromones exist. In the same region there is also a nerve called the *nervus terminalis*, which is considered to be a sensory nerve connecting with the olfactory bulb, but there is uncertainty regarding its peripheral distribution.

In humans the unmyelinated axon of the olfactory receptor cell extends proximally into the olfactory bulb through the cribriform plate, a part of the skull. There it enters the olfactory bulb and makes synaptic connection with other cells closely packed with other axons from neighboring cells. They come together in the so-called *lamina propria*, a layer of fibrous connective tissue, which also contains the Bowman's glands, located below the basal cells. Each receptor cell is assumed to function as an independent unit, because it makes no apparent physiologically active synaptic connection below the bulb. This is surprising, considering that the number and density of olfactory cells approaches that of visual receptors in the retina. The great number of olfactory receptors may preserve the ability to keep on responding under constant stimulation, and may counteract adaptation. There is no definite proof of this as yet, and no morphological evidence has been presented that these receptors differ from one another. Although they may vary between species, in any one animal, including humans, they look alike. Do they function differently electrophysiologically? No definite answer has been reached on this fundamental question of how the receptors "read" incoming information, but we must keep in mind that these receptors are continuously developing and that their response specificity may change in this regeneration process.

The Receptor Function

Two electrophysiological responses, both involving voltage transients, have been most frequently recorded from the epithelium (Shepherds, Getchell, & Kauer, 1975). One is the evoked action potential from the receptor cell of fiber. It is a so-called single-unit recording from an electrode placed in the epithelium either in the cell or externally. This provides basic data regarding evoked spikes to various odorants. (See Figure 2.3.) Although this potential should provide a complete description of the neurological basis for odor perception, the results have been complex—according to some investigators, chaotic (Gesteland, Lettvin, & Pitts, 1965). Although many researchers had hoped to find receptor specificity, the conclusion seems to be that receptor cells are broadly tuned, and each cell may respond to qualitatively different odorants. The effects of odorants on such

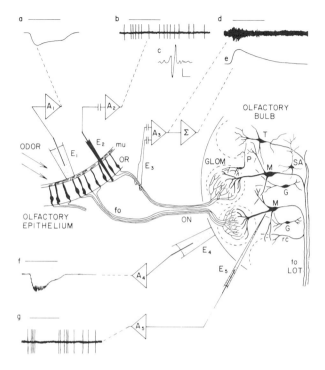

Figure 2.3. "A schematic representation of the cellular anatomy of the peripheral olfactory system of vertebrates and of its odor-evoked electrical activity. Olfactory receptor neurons (OR) lie in the olfactory epithelium, the cilia on their dendrites bathed by mucus (mu). Saline electrode E_1 measures the electro-olfactogram (shown in inset a). Electrode E_2, a metal microelectrode, measures extracellular spike activity in single receptor cells (shown in insets b and c). Bundles of axons form the fila olfactoria (fo), which coalesce as the olfactory nerve (ON). Bipolar wire electrodes E_3 record asynchronous spike activity from a bundle of axons or the whole nerve (shown in inset d). The signal can be rectified and summated with a decay (Σ) to give a smoothed measure of evoked activity (shown in inset e). Axon terminals branch at the olfactory bulb in spherical concentrations of neuropil called glomeruli (GLOM), where they synapse with processes of second-order neurons. Extrinsic neurons of the bulb include mitral cells (M) and tufted cell (T), whose axons project centrally as the lateral olfactory tract (LOT). Neurons intrinsic to the bulb are in periglomerular cells (P), which form lateral connections between glomeruli, granule cells (G) and short axon cells (SA). Mitral cells send recurrent collaterals (rc) toward the glomeruli. Electrode E_4 is a saline pipette which records bulb surface potentials (shown in inset f). A microelectrode, E_5, records extracellular activity of single second-order neurons (shown in inset g). . . . The lateral olfactory tract projects to the anterior olfactory nucleus (AON), olfactory tubercle (OT), prepyriform cortex (PP), amygdaloid complex (AM), and the transitional entorhinal cortex (TER). Projections from the olfactory tubercle, prepyriform cortex, and amygdaloid complex go to the hypothalamus (HYP). Other central connections and centrifugal pathways to the bulb have been omitted. The horizontal lines in insets a, b, d, f, and g represent the stimulus delivery period, 5 sec. All responses are drawn with positive voltages upward. Inset c shows the waveform of an extracellularly recorded receptor action potential seen with an expanded sweep. For this the calibration marks represent 200 μV and 5 msec." (Figure and figure legend from Gesteland, 1978, p. 263.)

cells may be ordered on a scale, with categories ranging from maximum excitation to maximum inhibition, but it is not clear what attribute of the odorant is correlated with this and how the information is coded. One of the most significant papers in olfaction is one by Gesteland, Lettvin, Pitts, and Rojas (1963) in which they first reported recording from single olfactory units, which are very small and tightly packed anatomical structures. Forty percent of the cells tested responded to 26 different odorants; no cell responded to all the odorants or showed any unique response to any one of them. In general, there seems to be a tendency for all cells to respond differentially to different odorants, but the specificity of receptors remains a problem.

The second kind of electrophysiological measure used is the electro-olfactogram (EOG), which is obtained by placing an electrode on the surface of the epithelium with the ground elsewhere on the body, thereby measuring receptor currents and not receptor action potentials. (See Figure 2.3, inset a.) This was another milestone in the field, first accomplished by Ottoson (1956). When an odorant is administered to the epithelial surface, a slow negative shift in voltage, assumed to be the summation of many generator potentials from single receptors, may be observed. (See Figure 2.3.) When the same procedure is applied to nearby epithelial surfaces that are innervated by neurons but not olfactory ones, no EOG is observed.

Although it is not clear how responses to single cells relate to the EOG, or how different odorants are discriminated, different patterns of responses have been obtained for qualitatively different odorants. Relatively different recordings were obtained by Mustaparta (1971) when the electrode was moved to different parts of the epithelium, which could be the result of different receptors in those locations. As far as response to intensity is concerned, Ottoson (1956) found that the amplitude of the negative voltage shift of the EOG obtained from the frog's epithelium grew as a positively accelerated (power) function of the odor concentration. The response of the receptors is converted directly into information about quality and intensity in the olfactory brain.

How does an increase in intensity of stimulation affect the single unit? In this case the relationship between concentration and the number of spikes or rate of action potentials from single receptors is apparently quite complex (Gesteland, 1978). Not all cells respond in the same fashion. One odorant may depress the response of one cell to a level below its spontaneous level, increase the level of another, and have no effect at all on a third. Regarding the intensity of stimulation, each cell seems to show the maximum number of spikes to a particular concentration. Fewer responses are evoked for either weaker or higher concentrations. This finding is interesting, because the EOG shows a monotonic change with concentration, and the question arises of what it reflects. The EOG has been assumed to code intensity directly, but it could be the sum of activity in all cells.

These are empirical reasons why no general conclusion can be drawn regarding how olfactory input is coded in vertebrates. A clearer picture is evident in insects, which to a large extent are controlled by odorants (Kaissling, 1971). Insects seem to have two fairly distinct classes of receptors. The *specialists* are olfactory receptor cells that respond to specific odorants such as sex attractants, and possibly to food and humidity. The *generalists* respond to flowery, aromatic, and other odorants and are characterized by overlapping spectra. (A *spectrum* is a description of a receptor in terms of its threshold of sensitivity of response to various odorants.) However, even this classification scheme is not airtight, for cells described as specialized generalists and generalized specialists have been reported (Kaissling, 1971, p. 388). In any case, the olfactory system of insects is quite different from that of vertebrates.

The problem of how odors are represented in the olfactory system of vertebrates will be discussed subsequently, but it should be noted here that part of the reason for this apparent problem of understanding how receptors function is that the molecular attribute that triggers the receptor is not known. It is this crucial problem about which so many theories have been proposed, as mentioned in Chapter 1. In fact, it is only recently that electrophysiologists have agreed that the EOG is a summation of many small generator potentials.

The Olfactory Bulb

According to Gesteland (1978), "lateral interactions, efferent feedback, summation, and reciprocal synapses complicate stimulus–response relationships at all levels of the olfactory brain higher than the receptors. The receptors therefore appear to be where the relations between neural activity and stimulus quality and intensity should be most transparent [p. 261]."

Others have nevertheless considered the possibility that the *ordering* of *odorants* may be more evident higher up in the olfactory system. Leveteau and MacLeod (1966) studied coding of responses by recording electrophysiological responses from the glomeruli, in which thousands of olfactory receptors may end in one synaptic region. They tested 128 glomeruli with 12 different odorants. The results indicated that each glomerulus would usually respond to only some and not all odorants. The important lesson from this study is that more order is evident in the sensory information at this point in the pathway. It is possible that we will learn how the information is transmitted at the periphery of the olfactory system and centrally without learning how odorants and receptors interact.

The main parts of the olfactory system are the olfactory mucosa, the olfactory bulb, and the olfactory brain. Most of the earlier electrophysiological work was done on the olfactory bulb, the work of Adrian (1954) represents a highlight. As the axons of the olfactory receptors leave the olfactory epithelium, they pass through holes in the cribriform plate in the skull and connect with the surface of the olfactory bulb, a cylindrical extension of the brain sketched in Figure 2.4. There is a topographical relationship between the olfactory epithelium and the bulb, such that the anterior and posterior receptors of the epithelium project to anterior and posterior parts of the bulb. At the same time, several hundred of the olfactory axons connect with a single cell in the bulb. These cells in turn are arranged in distinct concentric clusters or layers with connections to various other kinds of cells. It is here that the sensory information is processed via synaptic connections between these cells. Inhibition appears to be a basic aspect of the processing (Shepherd, 1976). A glomerulus seems to function as a unit when stimulated with odorants, such that all of its cells either respond or fail to respond to any one odorant. In the rabbit, more than 25,000 receptors project to each of approximately 2000 glomeruli, and from them the information is transmitted to about 45,000 mitral cells and 130,000 tufted cells (Allison & Warwick, 1949). These connections are complex, and not well understood.

It is the axons of the mitral cells that make up the olfactory tract, relaying the information to the brain after it has been processed in the bulb in ways not yet appreciated. Electrophysiological recording from these neurons show that they act spontaneously—that is, without stimulation from the

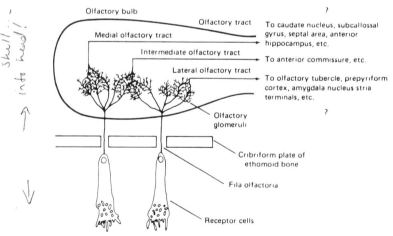

Figure 2.4. A schematic drawing of the olfactory bulb and ascending pathways. (Figure 3.47, p. 147 in *The Psychophysiology of Sensory Coding* by William R. Uttal. Copyright © 1973 by William R. Uttal. Reprinted by permission of Harper & Row, Publishers, Inc.)

olfactory epithelium—and, what is even more potentially important, that a particular odorant may excite some of these neurons and inhibit others. Specificity in terms of differential responsiveness of these neurons seems evident (Døving, 1964). Overall, about 60–70% of them seem to be inhibited by stimulation, 10% are not affected at all, and the rest are excited, whereas the receptors in the epithelium are usually excited and seldom inhibited.

The projections of the mitral cells and tufted cells are different. The axons of the mitral cells project to the hemisphere on the same side and the tract goes from the bulb to areas at the base of the brain, the olfactory tubercle, and other structures on the ipsilateral side of the brain. The axons of the tufted cells may synapse with cells in the anterior olfactory nucleus that connect with the contralateral olfactory bulb via the anterior commissure. This offers the possibility of projection to the other cerebral hemisphere and the other olfactory bulb. But there is little information about the synaptic connections beyond the bulb.

Although the olfactory receptors appear to function as a single unit, at the level of the bulb there are complex interactions between cells of the bulb as well as between the bulbs. This is undoubtedly an important but not yet understood aspect of processing olfactory information. Also, both bulbs may be influenced by other areas of the central nervous system with which the bulbs are interconnected. Apropos of this, if one bulb is stimulated electrically, it tends to suppress the activity in the other bulb, whether the activity is spontaneous or evoked by odorants. What is needed and now possible to obtain is precise quantitative stimulation and recording of various cells (not all of which have been mentioned here) and their connections, to understand the functional organization of the olfactory bulb and thus how odor quality may be coded (Shepherd, Getchell, & Kauer, 1975).

As we shall see later, since no cells seem to be activated by a single or "primary" odorant, it may alternatively be assumed that odor quality is coded by the pattern of activity in many cells. Odorant intensity may be reflected in temporal changes in the response of the single cells with characteristic increases and decreases in spikes for excitation and inhibition respectively (Gesteland, 1978). It seems important to explore the psychophysics and physiology of intensity systematically before attempting to clarify quality coding.

The Olfactory Brain

The olfactory bulb is actually the enlarged terminal part of the olfactory lobes, which are an extension of the front or anterior part of the cerebral hemispheres. The two olfactory nerves are the first cranial nerves that originate in the mucous membrane of the olfactory epithelium and lead to the area behind the olfactory bulb. This area is in turn connected to other parts of the brain, as mentioned earlier (MacLeod, 1971). However, although parts of this structure have an olfactory function, it does not serve only one function, which illustrates the complexity of the problem. The parts of the brain that seem primarily olfactory are the olfactory tubercle, the prepyriform cortex, part of the amygdaloid nuclei, and the nucleus of the stria terminalis, which cluster around the base of the brain at the end at that lateral olfactory tract. From the tubercle, prepyriform cortex, and amygdaloid cortex the projections go to the hypothalamus. The connections between the epithelium and the cortex are complex and by no means direct. Other structures are involved that also serve food intake, temperature regulation, sleeping cycles, and emotional behavior. It is interesting that the olfactory epithelium is only two synapses away from the hypothalamus; it is also interesting to speculate that this is consistent with the apparent importance of odor in reproductive behavior, at least in animals in which the vomeronasal organ may play a significant role. There are also connections between the olfactory brain and those parts of the brain serving vision, audition, and taste, as well as other areas more central to the brain. This makes it possible for other systems to influence olfaction by efferent connections with the bulb, in particular through inhibition. Long ago Lashley and Sperry (1943) showed that the rat's ability to discriminate between odors could be done with only the bulb intact. But as one moves up in the cortex, fewer and fewer neurons are excited, indicating that a more and more specific stimulus is required, analogous to the situation in the visual pathways (Døving, 1970). Other studies involving excision and stimulation of these neural structures have been made, but no clear picture has emerged. Nevertheless, W. J. Allen (1940, 1941) showed that severing the connection between the bulb and the pyriform–amygdaloid areas, or ablation of them, prevented classical conditioning of dogs to make different responses to different odors. However, they could still be taught to make simpler responses using an odorant as a cue. Research with monkeys has shown that the preorbital cortex is involved in discriminating the similarity of odorants (Tanabe, Iino, & Takagi, 1975; Tanabe, Yarita, Iino, Osshima, and Takagi, 1975). This may help explain the findings that aphasic humans with lesions in the frontotemporal lobes also have serious difficulty in processing information about odorants (Mair, Capra, McEntee, & Engen, 1980; Mair & Engen, 1976). Surgery to relieve epileptic seizures indicates that temporal lobe lesions involving the uncus, amygdala, hippocampus, hippocampal gyrus, and fusiform gyrus affect performance in detection and recognition of odors (Rausch & Serafetinides, 1975). There are both afferent

Coding

and efferent connections between these structures and the bulbs. Their role is probably complicated with regard to both the psychophysical task for the patient and a neurological explanation of the data.

Coding

There are two aspects of how perceived differences in intensity and quality are mediated by or represented in the nervous system. I have already referred to both problems, but I shall here discuss the theories that have been proposed to account for them.

Odor Intensity

The mechanisms that account for experienced odor intensity or strength are still unknown. There is evidence of a correlation between the amplitude of the EOG from the frog's epithelium and judgments by human subjects of the perceived intensity of the various concentrations of the same odorants. The investigators believe that the correlations are good enough for EOG records to be used as biological detectors of odor (Drake, Johansson, von Sydow, & Døving, 1969; von Sydow, 1968). In separate experiments, both the EOG and the human judgments of intensity were found to grow as power functions of the odorant concentrations determined from gas chromatographic analysis. As mentioned, it is assumed that the amplitude of the EOG is a function of currents from a number of receptor cells where the electrode is located. Therefore, it is possibly such information that is transmitted to the brain and there interpreted by the human subject as odor intensity.

However, the receptor cells vary in their sensitivity and may not respond to all odorants. In addition, we have also seen that the response of single cells grows in quite a different and nonmonotonic fashion with an increase in odorant concentration. There is first an increase in excitation, or the number of spikes evoked. At higher concentrations, inhibition may occur, with fewer spikes than would be obtained spontaneously, without any stimulation. Therefore, it is not at all clear how results from single cells correlate with those from the EOG recordings.

Odor Quality

In sensory physiology, intensity is usually found to be coded in terms of the number of spikes from each cell and the number of cells firing, both typically increasing with increased intensity of stimulation. I have already indicated that the search for the code according to which qualitative differences between odorants are discriminated has classically consisted of a search for specific receptors, each responding to a particular odorant or attribute of an odorous molecule. This problem has also been approached through the analysis of the similarity of odors. Agreement has been obtained between similarity judgments by humans and electrophysical responses from the frog to pairs of qualitatively different odorants, with similarity being scored as to whether or not cells in the bulb were excited, inhibited, or unaffected by a particular pair of odorants (Døving & Lange, 1967). An experiment involving homologous alcohols that have similar odors indicated that the pattern of electrophysiological responses from the frog is also more similar as the odorants become more physically similar (Døving, 1966). Some researchers believe that an odor classification of qualitative differences may emerge from multidimensional analysis of such similarity results (Døving, 1970; Erickson & Schiffmann, 1975).

However, one need not yet give up on the idea of specific receptors, especially as it might involve the effect of inhibition, as in the case of the results of Gesteland (1978) referred to earlier. One reason for the apparent chaos in the findings regarding specificity of receptors with respect to quality may be the confounding of quality and intensity; intensity is usually not evaluated carefully psychophysically in electrophysiological studies. One cannot be sure that the olfactory system works on the simple kind of frequency code indicated earlier.

Amoore's (1970) stereochemical classification system seems to be the most popular model proposed for describing odor quality and similarity. The idea was first suggested by Moncrieff and it proposes that odorants having similar odors also have similar molecular sizes and shapes—a descendant of the atom theory mentioned in Chapter 1. Lucretius believed that pungent odors were associated with hooked or barbed molecules, whereas sweet-smelling odors were associated with smooth molecules (Davies, 1971).

Amoore first proposed seven such stereochemical classes, each corresponding to a primary odor (ethereal, camphoraceous, musky, floral, minty, pungent, and putrid). Subsequent research, however, has suggested that more classes are required to account for all the different odors experienced by humans. There has been a search for cases of specific anosmia, the inability of a person to smell a certain odorant, which Amoore (1971) believes may be genetic in origin and may involve primary odors and specific receptors. This is of main interest in the present context—namely, the part of the theory that states that on the olfactory receptors there must be sites that will accommodate the different shapes. It may be described as a *ball-and-socket theory*. (See Figure 2.5.) For example, camphoraceous

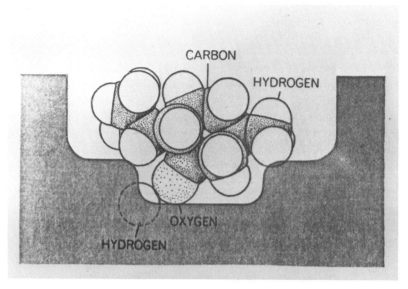

Figure 2.5. A drawing of the molecule of *l*-menthone, which elicits a pepperminty odor, showing how the molecule fits into a special cavity, according to Amoore's theory. (From "The Stereochemical Theory of Odor," by J. E. Amoore, J. W. Johnston, Jr., and M. Rubin. Copyright © by Scientific American, Inc. All rights reserved.)

odors all have molecules shaped like a ball, according to Amoore's research, and there should be a bowl-like site on some receptors to accommodate it. An ethereal odor is shaped more like a sausage and would need an oblong, dishlike site to accommodate it.

Amoore does not attempt to describe how the molecule in such a site might generate a receptor potential. From what was said earlier, it should be clear that there is no evidence that such sites exist or even that it is a plausible idea. Yet, even though such sites may not be found, and even though it might not be possible to sort odorants into specific classes, the shapes of molecules may still correlate with odor quality. Odorants with similar stereochemical molecules may produce similar odors. Even before one gets this far, however, there is the fundamental psychological problem of having people agree on how to label their experiences. After hundreds of years of research on odor classification there seems little hope of achieving agreement that there are such primary and special odor qualities as "ethereal" or "camphoraceous."

There are two other atomic theories that are often referred to in the literature. One is the *penetration-and-puncturing theory* (Davies, 1971). Odor sensation is thought to be caused when odorant molecules puncture the olfactory receptor cell, which then causes sodium and potassium to flow across the cell membrane and initiate an action potential. It is proposed that there may be thousands of such accessible sites on each cell. The intensity of the odorant will affect the number of punctures per cell by a certain odorant, whereas the quality of the odorant depends on the rate at which the holes left in the sites are healed. This in turn depends on the desorption rate and molecular size and geometry of the molecule, and it is assumed that these characteristics will denote uniquely the odor of any odorant and also that they will predict shades of differences between them. This theory therefore deals with dimensions rather than classes of primary odor, as in the stereochemical theory. Similarity of perception can be dealt with nicely in the case of color, which is correlated with wavelength of light. Unfortunately for the experimentalist, there is no such possible stimulus manipulation in the study of olfaction. One cannot change frequency and keep intensity constant, as with light. Changing the odorant may involve complex changes in both quality and intensity as defined by this theory. Not enough is known about the odorants to make such systematic changes for such a study (Beets, 1970). The chain length of alcohols mentioned previously do predict similarity of odor perception to some extent, but one cannot generally predict odor similarity from physical or chemical similarity of the odorant.

The next theory of this type proposes that the vibration frequency of the odorant molecule is the attribute that affects the receptor cells (Wright, 1977); that is, the atoms of a molecule are not fixed but move at predictable speeds. These speeds may be different for different odorants. There may be as many as 25 primary odors mediated by different receptors and tuned to different frequencies. These frequencies can be measured by the infrared or Raman spectrum of the odorant. The theory has come in for especially severe criticism, because some odors of different quality have similar frequencies (see Moulton & Beidler, 1967).

All three theories mentioned suffer from the same basic lack of information regarding the receptor cell and the cilia with which the molecules must interact in a chemical or physical way. Although no specific receptors have been found, the theories cannot, however, be rejected, because there must be some kind of molecule–receptor interaction in olfaction as in other modalities. Also, commercially produced insect traps have resulted from predictions of "favorable" versus "adverse" frequencies of olfactory quality based on infrared spectra of the odorants (Wright, 1974). The idea of specific receptors is not dead, just unsupported. Recall that Mustaparta (1971) obtained different EOG records from different parts of the epithelium, suggesting the existence of different receptors. Although they are not necessarily mutually exclusive regarding how stimulus properties affect cells, some alternatives to the theories just discussed are broadly described as pattern theories (Moulton, 1976b). However, in the codfish, discrete olfactory bundles are associated with different behavior, and each bundle is apparently stimulated by a specific odorant, thus indicating a spatial basis for odor discrimination (Døving & Selset, 1980).

Pattern Theories

There are at least two pattern theories considered at present. Both of them assume that there are only generalist receptors. Although each generalist may respond to a wide range of odorants, it may have its own individual spectrum; in other words each receptor may be broadly tuned. The response from a number of such generalists may together form a unique response matrix for each qualitatively different odor. Any quality then reflects information carried by two or more receptors. This kind of pattern theory is often described as *cross-fiber pattern* (Erickson, 1968). For example, receptor 1 may with its spectrum be more responsive to odorant A, and receptor 2 to odorant B. Both receptors respond to both odorants, and recording from either one would reveal that. The greater the number of odorants and receptors, the greater the apparent chaos, if the goal is finding specifically tuned receptors. If one observes one receptor cell at a time, it may be found more sensitive to some substances than to others, but whether one quality or the other is experienced may only be known by observing both receptors simultaneously. We might then find that for odorant A, for example, there is more activity in cell 1 than in cell 2, whereas for odorant B the reverse is the case. There is really no such information available in the sense of smell, but such a theory has been shown to be plausible for other modalities, including taste.

The other pattern theory, though not necessarily contradicting the cross-fiber pattern theory, has a longer history and some actual data supporting it. This is the *gas chromatographic model* (Mozell, 1966) based on earlier work by Adrian (1950; 1953). According to this model, different odorants may be maximally effective at different places along the epithelium from front to back; in addition, their molecules spread across the epithelium at different rates.

One might call this the *time-and-place theory*. The spatial–temporal factors could depend on different binding strengths with which different molecules are adsorbed to receptor cells. Since the olfactory system is spatially represented in the bulb there is anatomical justification for the theory. The medial branch of the olfactory nerve serves the external part of the naris where the odorous air first enters the olfactory system, and the lateral branch serves the internal naris where air exits. Mozell (1966)

showed that the ratio of activity in those two branches varied for different odorants when tested at different concentrations, both for the magnitude of the response and its latency. Support for this spatial–temporal pattern was reinforced by showing that reversing the flow of odorous air from interior to exterior naris also reversed the results from the two branches (Mozell, 1970). Substituting the frog's olfactory system for the column of a standard gas chromatograph showed that it gave analogous results and may thus be considered a model of the olfactory system (Mozell & Jagodowicz, 1973). Earlier Beidler (1957) had suggested that molecules that are relatively strongly adsorbed to the olfactory epithelium may elicit smaller differences in response from the two nerve branches than molecules of smaller affinity with the receptor cells. In fact, the retention times in the column of a gas chromatograph predict rather well the retention times for more than a dozen diverse odorants obtained from the frog's epithelium. Thus, molecules moving across the epithelium at different rates may be the first stimulus event in odor perception, and there is evidence that it takes place in a spatial–temporal pattern.

These observations also call attention to the mucus, the medium in which the stimulation takes place. In addition to the receptor and cilia, the mucus probably plays a role in coding odors. For example, it has been suggested that the change in odor sensitivity observed during the menstrual cycle is related to hormonal effects on the thickness of the mucus layer (Mair, Bouffard, Engen, & Morton, 1978). This cyclic change in odor sensitivity had originally been thought to hold for only certain biologically significant musk-like odors (e.g., Exaltolide) and thus to be an indication of the possible existence of human pheromones. A more traditional and simpler explanation in terms of the physics of the transduction of a stimulus into a neurological signal is suggested in Chapter 7.

chapter 3
PSYCHOPHYSICS

Psychophysics is the branch of psychology that describes in quantitative terms the relationship between physical stimuli and psychological responses. It is primarily concerned with perceived intensity as a function of concentration. Since olfaction is classified as one of the chemical senses, one might expect the term *psychochemistry* to be used, and one might also expect the existence of a chemical reaction in mucus initiating a neural response resulting in a sensation. But as was pointed out earlier, the salient chemical attributes in olfaction have yet to be determined. In fact, it is possible that physical attributes (for example, intramolecular vibrations) are the crucial factor differentiating those compounds that are perceived by the sense of smell from those not perceived.

In his well-known book on the chemical senses, Moncrieff (1967, pp. 597–600) describes all the features of odorants that researchers have believed to be olfactory stimuli, and then groups the hypotheses according to four main factors: (a) volatility (to move molecules to the olfactory epithelium); (b) solubility in lipids, but also in water and protein; (c) chemical reactivity in one form or another, not yet understood; (d) molecular vibrations; and (e) adsorbability. Although none can be considered the main salient characteristic of the stimulus, all may be involved because perception of odor may involve several processes, some chemical, some physical, in the olfactory mucosa.

At equal concentrations, some odors are stronger than others. For example, the threshold for ethyl mercaptan is about .00000066 mg/liter, compared with 4.5 mg/liter for carbon tetrachloride (Wenger, Jones, &

Jones, 1956). People are so sensitive to the mercaptan that it can be added to odorless but toxic gas to warn them of its presence. It is not known why mercaptan is so potent, and that is part of the fundamental problem of defining what the stimulus is.

Two contemporary hypotheses should be added to Moncrieff's list. According to Beets (1964), the perceived intensity of an odorant depends on chemical affinity with the receptors, which is determined by the "profile" functional group to which it belongs, whereas quality depends on the physical profile of the odorant molecule, an idea similar to Amoore's stereochemical theory (Beets, 1968). A more complex and less traditional hypothesis is that the perceived intensity of an odorant is the result of the combined effect of its molar volume, hydrogen bonding, and polarizability (Laffort, 1969).

All of the hypotheses alluded to can only be considered tentative, but progress has been made in reducing the number of them. Neither the intensity, nor the quality of an odorant can be specified physiochemically at this time; several chemical and physical parameters may be involved. To a first approximation, practical methods used in controlling odor must rely on the estimation of concentration in terms of the number of molecules reaching the olfactory epithelium, applying the technology called *olfactometry*.

Olfactometry

An *olfactometer* is an instrument designed to control and manipulate the concentration of odorants. Most of the work in olfactometry has involved the specification and comparison of thresholds for different substances and subjects, but interest and research in the response to suprathreshold odors have increased dramatically. The history of the principles of olfactometry, covering the 100 years prior to knowledge and application of gas–liquid chromatography, is presented by Johnston (1967) and by Wenzel (1948).

Zwaardemaker's Olfactometer

Zwaardemaker's olfactometer (Figure 3.1a) may now be of historic interest only. It consisted of two tubes, one hollow one sliding over another, which was open at both ends. The inner surface of the outer tube was covered with an odorant. By varying the overlap of the two tubes, the amount of odorant exposed could be varied. One end of the inner glass tube had a suitable nosepiece through which the observer could sniff. The technique was simple and produced results, but it is debatable whether the odorant concentration was proportional to the area thus exposed (Wenzel, 1948).

The Blast-Injection Olfactometer

Another criticism of Zwaardemaker's method was its reliance on the natural sniff, which might vary from trial to trial and from person to person and thus vary odorant flow rate and strength. One early innovation for better stimulus control involved presenting a measured volume of air to the subject by using a syringe (Elsberg & Levy, 1935). Elsberg's blast-injection olfactometer, as it is generally known, consists of a bottle containing a prescribed amount of a liquid odorant, which is allowed to volatilize until it reaches equilibrium with the air in the bottle. At that time pressure can be built up with a syringe and released by opening a stopcock attached to a nosepiece (Figure 3.1b). A certain amount of this vapor can then be delivered to the observer at a certain force with tubes and nosepieces. In addition to the presentation of a precise volume and concentration at a certain flow rate, this method was designed to overcome the variability expected from natural sniffing. The blast-injection method may still be in use medically for controlling the location of stimulation (Schneider, Costiloe, Vega, & Wolf, 1963). The syringe is also used in the method proposed by the American Society for Testing and Materials (Sullivan & Leonardos, 1974). However, its validity is suspect, because F. N. Jones (1955) has shown that observers may confuse the sensation of odor of the vapor with the sensation of touch of the blast (see also Benignus & Prah, 1980). This is a serious practical problem because it is difficult to discriminate the sensory quality of weak sensations. Another criticism of this kind of technique involves the lack of purity of the ambient air typically used to dilute the odorant (Wenzel, 1948). On balance, then, not much is left of the advantage of blast injection over allowing the subject to sniff the odorant.

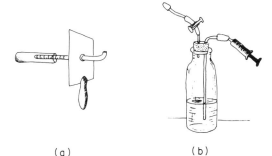

Figure 3.1. Drawings of (*a*) Zwaardemaker's and (*b*) Elsberg's olfactometers.

(a)   (b)

Sniff Bottles

Probably the most common method of presenting odorants involves placing them in tubes or beakers, so-called sniff bottles, and letting the observer sniff from them. Besides the problems already noted, the dynamics of liquids in such containers may produce variability. The vapors from sniff bottles are therefore always suspected of failure to conform to Raoult's law of vapor pressure (Haring, 1974).

Sniffstrips and equilibrium sniffers are used to provide a more constant vapor than sniff bottles and apparently more reliable response (Engen, 1964). Sniffstrips, a favorite of perfumers, are strips of blotting paper dipped in liquid solutions. One common variation uses a cotton swab, a wad of cotton wrapped around a glass rod (Engen, 1964). The equilibrium sniffer (Figure 3.2), suggested by Turk (1964) and used successfully in a number of experiments (Engen, Kilduff, & Rummo, 1975; Mair et al., 1980), consists of a fritted glass disk on top of a bottle containing the odorant. The bottle is kept stoppered till the vapor inside reaches equilibrium. When the stopper is removed for sniffing, the vapor pressure through the interstices of the fritted glass will be uniform. If one is careful to calibrate the filters one can compare different concentrations and substances in different bottles.

Another simple and sound approach is to use a glass vessel with a perforated filter disk, in the middle of which liquid odorant is kept in a wad of cotton. The vessel contains two ports, one above the filter and one below, and the observer sniffs through the lower port (Cain, 1977a). Finally, microencapsulation of liquid odorants in plastic capsules on paper from which the odorant can be scratched loose has also been attempted (Davis, 1979a).

All of these variations on the sniff bottle depend on natural sniffing. Although observers can learn to do this consistently and thus control the flow rate (see Chapter 2) and the volume of air inhaled, it is difficult, especially in a threshold experiment, for observers to resist a natural tendency to sniff harder when an odorant is weak. To overcome this potential problem Köster (1971) cleverly used a wedge as a stopper in the tubing between the odorant source and the observer's nosepiece. When the subject sniffs at a moderate strength, the wedge is lifted off its seat, allowing the vapor to flow around it. However, should the subject sniff too vigorously, the wedge will block the tube at the other end.

The simplicity of sniff bottles means flexibility and ease in manipulating the quality of odorants by presenting different odorants in different bottles. Intensity is varied by diluting the substance in a diluent appropriate to its solubility, such as water, mineral oil, benzyl benzoate, or diethyl phthalate. Different concentrations are typically stepped off geometrically by succes-

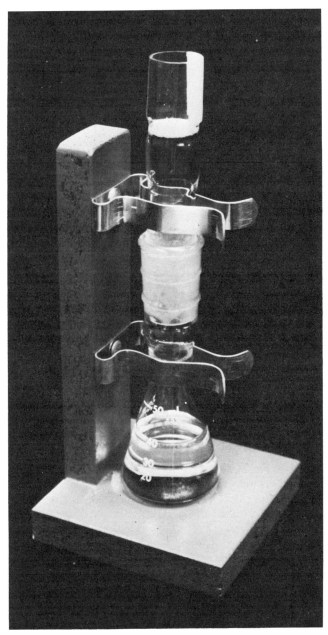

Figure 3.2. An equilibrium sniffer.

sively halving the concentration of the undiluted substance until one reaches a level at which the subject can no longer detect the odorant. Such a concentration series can be used for comparing thresholds, for scaling the perceived intensity of higher concentrations, and for obtaining ratings of the pleasantness of different substances or concentrations. For many purposes, sniff bottles, especially the modified types, are adequate. The essential limitation is that they do not readily lend themselves to precise determination of the concentration of odorant molecules in the vapor from the cotton or blotter. Generally, dilution in air or another carrier gas, such as nitrogen, is a more refined and sophisticated approach, and that involves olfactometry in the usual sense of that word.

Modern Olfactometers

Modern olfactometers (Figure 3.3) present carefully measured quantities of an odorant, in terms of both quality and intensity. Samples of vapors delivered are usually evaluated gas chromatographically for both the purity of the quality and the concentration (see, e.g., Eyferth & Kruger, 1970). Typically these olfactometers provide the observer with a constant gentle flow of vapor whose rate closely matches that of normal breathing. The vapor is kept near body temperature and is presented to the observer through a nosepiece or funnel. Although nitrogen is used by some researchers as a carrier gas, most use air. The air used to dilute the odorant is first purified and then saturated with the odorant, usually by flowing it through a bottle containing the liquid odorant.

After this air has been saturated with the odorant, it can then be diluted further by mixing it with purified air from another channel using a sparger to provide a good distribution of the odorant molecules in the air. There are two common methods, each with its supporters, for controlling the concentration of the odorant. One way is to flow the odorant through capillaries, each one calibrated for a desired concentration. The other way is to measure the odorant with a flow meter, a more flexible but perhaps less reliable method because of variability in human adjustments of the flow meters.

Olfactometers vary from relatively simple ones with one channel for testing one odorant at a time, with clean-up between use, to multichannel programmable ones with continuous gas chromographic monitoring of the vapor from many different odorants (Johnston, 1967). Some olfactometers are designed specially for animal psychophysics—for example, for fish (Døving & Schieldrop, 1975) and dogs (Marshall & Moulton, 1981). For environmental work a multichannel olfactometer can be put in a bus, trailer, or "sniffmobile" (Springer, 1974). In Sweden air pollution has been investigated with such a mobile laboratory (Figure 3.4) for over a decade

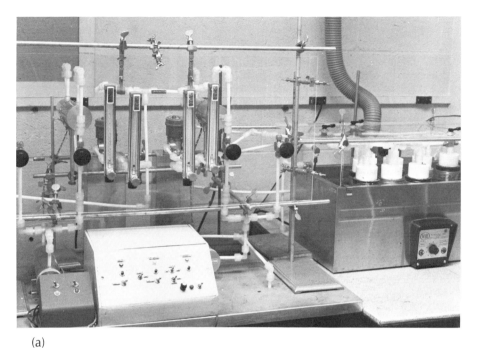

Figure 3.3. A photo of (*a*) an air-dilution olfactometer at Brown University and (*b*) the adjacent subject's booth with two sniffing ports. (After W. S. Cain, 1968.)

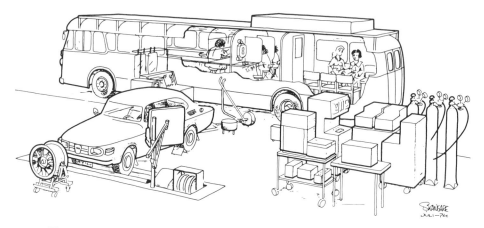

Figure 3.4. A mobile laboratory used in psychophysical analysis of automobile exhaust in Sweden. The exhaust is obtained directly from an automobile and delivered to the subject in an exposure hood, two of which are shown in the middle compartment of the bus (see also Figure 4.2). The front compartment of the laboratory houses the experimenter and the control apparatus, and the rear compartment is used as an air-conditioned waiting room having constant temperature and humidity. (From Berglund, Berglund, & Lindvall, 1977.)

(Lindvall, 1970a). Samples of air in any neighborhood or building can be collected, analyzed physiochemically on the spot, and presented to human observers for their evaluation of intensity and quality. When odor is judged to constitute an annoyance or a health hazard, odor-control technologists remove the odor or dilute it to safe or hedonically acceptable levels (see also Duffee, Jann, Flesh, & Cain, 1980).

Odor-Control Technology

Perception of odor does play a significant part in defining air pollution. For example, about 50% of the complaints about air pollution involve odors (Feldstein, Levaggi, & Thuillier, 1974). There has even been serious research on how neighborhood odors influence the amount of money people are willing to pay to live in an odorfree neighborhood (Flesh, 1974). Some odorants can be sampled from the air and measured physically, but for the most part one must rely on human judgments of intensity and unpleasantness. Knowledge of psychophysics and human engineering is indispensable in this field.

Food, perfumes, and flowers generally have pleasing odors but even they may become obnoxious at the high concentrations experienced by those living near a perfume manufacturer, or a coffee-roasting plant, for example. However, the sources of most annoying odors are factories producing paper, petroleum, phthalic anhydride, fertilizer, roofing material, soap, coke, metals, resins, rubber, adhesives, pesticides, pharmaceuticals, paint, essential oils for food and perfume, textiles, and leather, and of course gasoline and diesel engines, dumps, agriculture, and sewage plants. Still, although many of the sources of odorous pollutants are known, the reason why the odor is unpleasant is not at all understood. For example, "the compounds that comprise diesel exhaust odor are so numerous and complex that they defy their complete classification and direct measurement by chemical–instrumental methods [Springer & Stahman, 1974, p. 409]." Odorants can be reduced by processing them through a wet scrubber, by dilution, by conversion of the odorant to nonodorous substances (usually through oxidation to carbon dioxide and water) by adsorption, by a change in the production process causing the odor, or by masking or attenuating the malodor by adding another odorant to it (Haring, Turk, & Okey, 1974).

Emission Stacks

Diluting the strength of the odor by using tall emission stacks to reduce the concentration reaching human beings is a popular approach to reducing odor. A relatively small increase in stack height causes a relatively large attenuation of odor (Hardison & Steenberg, 1974). If this does not work for large volumes of easily detected pollutants, one or more of the other modern chemical methods must be used. Typically a combination of these methods is needed for any one source.

Scrubbers

Another well-known method is adsorption by the use of activated carbon, as in air-conditioning systems. A variety of wet scrubbers, which may also involve adsorption, are often used. The vapor is directed through a venturi or tower containing various packings to mediate transfer of the pollutant into the liquid phase, where it can be exposed to a solvent such as sodium hydroxide for mercaptans, hydrochloric acid for amines, and potassium permanganate for aldehydes (Dickerson & Murphy, 1974). Wet scrubbers can also be used for particulates that can readily be carried away with the wind. Serious attention is being paid to the complex effect of methodology in this regard. One way to control such sources of odor is incineration or combustion, thereby converting the particulate to carbon dioxide and water before it gets away (Hellman & Taylor, 1974).

Masking

The most psychological and possibly most effective approach to controlling the perception of odors is masking, or counteracting a malodor by introducing another pleasant odor. Masking and counteraction involve dilution of the concentration of any one odorant with another. By mixing them one takes advantage of the fact that on the psychological scale of intensity odors do not sum arithmetically when one is added to another. If the subjective strength of odorant A is 2 and the strength of odorant B is 3, the subjective sum is not 5 but less, depending on the nature of A and B. Zwaardemaker's findings on compensation (described in Chapter 1) meant that the intensity of the mixture was less than either A or B; that is, less than 2 in the present example. A more typical finding, however, is described as compromise, in which case the mixture would smell stronger than A but weaker than B. Although this method may work well for weak odors, its effectiveness, unfortunately, is inversely proportional to the potency of the malodor. (See Chapter 5.)

As in other areas of science and engineering, rapid progress has been made during the last decade or two in the development of techniques of odor control. But even this odor-control problem is affected by the lack of understanding of the odor stimulus. The effectiveness of the techniques, therefore, must be evaluated by human psychophysical observers rather than by physical instruments. Having examined how odors are produced and how they are controlled by manipulating physiochemical correlates of odors, we will now explore how the intensity of the odor the observer perceives varies with the magnitude of these physical correlates.

Odor Classification

"One of the pleasures of writing," according to Saul Bellow, "is being able to deal in certain primitive kinds of knowledge banished from ordinary discourse, like the knowledge brought by smells. I must be a great smell classifier." This quote is from an interview by Jane Howard (1970), who then illustrates the evocativeness of Bellow's writing with the following samples: "the damp stench of the subway," "the odor of tomatoes as they burst on the vine," and "the multitude of scents generated by women."

Adjectives and Primary Categories

The most primitive form of quantification is nominal scaling (S. S. Stevens, 1951). Applied to our problem, it involves the ability to name an odor and group it with other, similar odors. The only rule is that only one

Odor Classification

label, whether it be an adjective or number, can be applied to one class. In odor classification the typical method has been to use adjectives such as *goaty* and *minty* to name "primary" odors, analogous to the naming of primary colors of blue, green, and red. The real value of these primary colors is not so much the specific colors or labels themselves but that they can be defined in terms of the physical correlate of wavelength, such as 460, 530, and 640 μm. By mixing these wavelengths one can produce any other color of the visual spectrum. Color is therefore the ideal psychophysical model for other modalities. The question then is whether *goaty* and *minty* or other categories analogous to blue, green, and red do exist. If they do, one can begin the fundamental search for the physiochemical correlates analogous to wavelength, and then to how they affect the olfactory epithelium psychophysiologically.

As has been mentioned, no such correlates have been isolated. Even though odor experience must be the result of a reaction started when molecules interact with the epithelium, it has not even been possible to pin down the physiochemical attribute that serves as a basis for differentiating between compounds that do and do not activate this sense, let alone finer discrimination between such categories as goaty and minty odors. From the beginning the odor problem was more complex or multidimensional than the color problem. Failing to solve the problem by manipulating chemical and physical variables, researchers have tried to find the answer by examining odorants judged to have similar odors, to determine whether they do have a common physical or chemical attribute. A nice summary of attempts to describe odors with adjectives and relate each adjective to a group of compounds has been provided by Boelens (1974) and is reproduced in Table 3.1.

Harper *et al.* (1968) present a thorough review of attempts made to classify odors. The first such attempt was the work of the botanist Linnaeus in 1756. He proposed that there were seven such classes: aromatic, fragrant, ambrosial (musky), alliaceous (garlicky), hircine (goaty), repulsive or foul, and nauseating. Linnaeus then divided these seven classes into four major groups based on their hedonic or affective aspect—for instance, fragrant and aromatic, versus foul and nauseating. Amoore, whose work was discussed earlier (Chapter 2), also has seven classes, which he calls ethereal, floral, pepperminty, camphoraceous, musky, pungent, and putrid. Amoore considered the last two to be special and nonprimary categories, associated with electric charges rather than stereochemical attributes of the odorants.

The number seven is close to the mode for the multitude of classification systems proposed since Linnaeus. At the low end are Crocker and Henderson's (1927) four classes: fragrant, burnt, caprylic (animal- or barnlike),

Table 3.1
Relationship of Chemical Structure and Perceived Odor[a]

Odor class	Chemical compound
Ethereal	Alkanes, alkenes, alkynes, alkyl halides, and nitro alkanes to about C_5; alcohols to C_3; ethers (linearly) to about C_8
Alliaceous	Straight-chain thiols, thioethers, thials, thioketones, and thioesters; linear dialkyl-, dialkenyl-, di-, and trisulfides
Green	Unsaturated linear alcohols, aldehydes, and esters from about C_5 to C_{10}; esters from about C_2 to C_{10}; linear ketones from C_5 to C_{10}
Rancid (fatty)	Saturated linear aldehydes from C_5 to C_{15}; methyl ketones from C_{10} to C_{15}; fatty acids from C_4 to C_{15}; alcohols and esters from C_{10} to C_{15}
Burnt (pyrotic)	Benzenoid hydrocarbons; phenols, cresols, xylenols, and lower ethers; substituted dioxofurons and pyrans
Aromatic (spicy)	Substituted benzoid and alicyclic (medium-ring) derivatives, with these substituents: hydrozyl, methozyl, dioxymethylenyl, carbonyl, allyl, and propenyl
Floral (fragrant)	Substituted (medium-ring) cyclic and isosteric compounds, with a carbon chain from about C_2 to C_8 and these functional groups: alcohol (esters), carbonyl, and carboxy (esters)
Woody	Two- or three-ring systems and isosteric compounds with about 12 to 17 C atoms and these functional groups: alcohol (ester), carbonyl
Musky	Macrocyclic (C_{14} to C_{18}) and isosteric compounds with cross-sectional area of 50 Ångstroms and a length-over-breadth ratio of 3.0; molecular weight of about 250, and well-exposed functional groups: alcohol, ether, carbonyl ester, and nitro
Nauseous	Pyrrol, pyridine, chinoline, indole and lower homologues, steroids and isosteric compounds; organic diamino- and aminosulfide compounds

[a] From Boelens (1974).

and acid. Harper *et al.* (1968) have the record with 44. A large number of classes usually results from attempts to resolve disagreements between different systems or to accommodate criticism of any one system. In color perception, it is not so much that the labels *red, blue,* and *green* are specific and indispensable as names of classes or primaries as it is the fact that understanding the wavelength makes it possible to manipulate and predict perceived color. That is, it is the selection of the three separate wavelengths that is crucial, not the particular color experience or color name. Blue-green, violet, and orange or wavelengths 410, 500, and 630, respectively, would do as well. The constraint is on the selection of wavelengths, not the names of colors. The wavelengths must be spaced on the spectrum such that by mixing them it is possible to obtain any other color, except those corresponding to these particular wavelengths (that is, blue-green, violet, and orange in the present example). No comparable psychophysical relations are known in case of odor perception.

The problem of describing odors can also be seen when one attempts to translate adjectives from one language to another. Even when narrowed down to English, it turns out that though *goaty* and *sulfurous* describe similar odors in the United States, in Britain they represent different types of odors (Harper *et al.*, 1968). This may be akin to observations that different languages and cultures divide the visual spectrum, which contains no obvious perceptual boundaries, into arbitrary categories described by color names. In English there are six categories called *purple, blue, green, yellow, orange,* and *red.* However, in Iakuti there is a single name for blues and greens (Brown, 1958, p. 238). It seems to be generally true that people have difficulty recognizing colors for which they have no good names (see Bolinger, 1968, p. 256). People are also generally poor at identifying odors of common household items (Engen & Ross, 1973). In any experiment up to 40 adjectives may be used to describe 50 diverse odorants. Varying the sample of odorants also has far-reaching effects. In a study of 24 odorants described as putrid in the context of diverse odorants, as many as 9 subclasses could be applied when they were judged by themselves (Johnston, 1968). The number of classes is a function of the context or diversity of odorants in a set presented for judgment. Although there may have been some progress in developing an odor glossary, more than 200 years of efforts by botanists, psychologists, and perfumers have not produced much, if any, usable and enduring knowledge.

Multidimensional Analysis of Similarity

To get around the linguistic problem of analyzing the words people use to describe their odor sensations, modern multidimensional scaling methods have been recommended (Bienfang, 1941; Davis, 1979b; Lange, 1970;

Schiffman, 1974a,b; Schiffman, Robinson, & Erickson, 1977; Wender, 1968; Woskow, 1968; Yoshida, 1964a, 1964b). These methods do not ask the observers for qualitative descriptions but ask them only to rate the similarity of odors. In a sense, the purpose is to work backward from similarity judgments through the odor space to qualities. It is hypothesized that there is an odor space in the observer's perceptual system that determines these similarity judgments. This subjective space has axes along which the odors vary in perceptual quality. The judged similarity of a pair of odors, for example, would be related to the proximity of the two odors in the space. Various judgmental methods may be used to define the similarity of the odors, and the judgments are translated mathematically by factor analysis to distances (Engen, 1971a). For example, if a pair of odors are judged to be highly similar, there must be a short distance between them. If they are judged very different, the distance must be great. There is no single way to convert such judgments into distances in space, and so there are also various methods that may be applied for that purpose, depending on the assumptions made about the nature of the space and the quantitative characteristics of the data. Next, since the odor space is not known, various mathematical models complete with computer programs have been proposed for describing what it might be like (P. E. Green & V. R. Rao, 1972; Romney, Shepard, & Nerlove, 1972; Yoshida, 1977). Bienfang (1941) proposed that the odor space is a sphere with an axis of clarity, a radius of strength, and a circumference of note. This model is analogous to the color solid with dimensions of brightness, saturation, and hue, respectively.

When the analysis is completed and the decision made regarding how many axes are required to describe the hypothesized space—no easy task with variable human judgment—one then has another difficult task, deciding how they should be named. Multidimensional scaling therefore does not by itself identify odor qualities. For the most part the naming has been done on a subjective basis, as Bienfang's effort illustrates. Given the description of the odor space, would one agree with him that the dimensions should be named *clarity, strength,* and *note*? We are back where we started, trying to solve a semantic problem. One dimension (or is it two?) seems to stand out; namely, the pleasantness and unpleasantness of odors. But it is not clear whether they lie on opposite sides of a common hedonic dimension or actually are psychologically different. It cannot be claimed that there is any better agreement on the nature or number of other dimensions that can be extracted from such data. However, the most basic problem concerns how one would compare such dimensions with neurological events, and here multidimensional scalers have nothing more to offer than odor classifiers.

But there are problems to be faced even before one gets this far. Multidimensional scaling and analysis generally assume that there are reliable and fixed subjective dimensions that exist in each individual, like other psychological dimensions such as attributes of intelligence. In the case of odors, such dimensions are implicitly assumed to be expressed in the way in which odor quality is coded by the olfactory system, whether by specific receptors or patterns of excitation of different receptors. These hypothetical dimensions are assumed to be applied by the individual in judging each odor. However, there are reasons to believe that a more dynamic and cognitive model of judgment applies to reality. For example, in the case of putrid odors, we saw that the judged degree of similarity is affected by context. The variability of the similarity of putrid odors is greater when compared among themselves than when compared with more diverse odors (Johnston, 1968). The more similar the odors, the finer the discrimination the observer would make. Implicitly, this amounts to using less abstract terms to describe their quality, which in turn must affect the dimensionality of the odor space. In other words, one might say that the more similar the odors in a set, the more analytic the observer becomes. Although the primitive input at the olfactory receptors may be classified, the processing of the information higher up in the brain seems to play a decisive role in "the knowledge brought by smells."

At this higher level, individual differences in judging odors may also be especially great because of unique experiences and special associations with odors (Engen, 1974). Although multidimensional scaling assumes that physiochemical actions and receptor functions will determine the outcome of the scaling and analysis, psychological factors may be decisive. The larger their role, the less valid the multidimensional approach to the odor-quality problem. One study did in fact find evidence for the existence of individual odor spaces, and concluded that individual differences were too large to be averaged into a meaningful general odor space applying to the whole group (Berglund, Berglund, Engen, & Ekman, 1973). It does not surprise anyone that such individual differences exist, but they are not usually reported because the purpose of such research is to pin down *the* odor space.

In another multidimensional experiment (Gregson & Mitchell, 1974), which is quite relevant here, similarity judgments were obtained for pairs of odors (e.g., odors of isoamyl acetate and methyl propionate), and also for pairs of odorants and common odor descriptors (e.g., *like mothballs* and *aromatic*). The former case is the typical task in multidimensional scaling; in the latter, the subjects were required to compare a real odor with an imaginary one. It was found that the presence of the latter kind of pair influenced the odor space derived for the real odor pairs. It is not clear whether this happened because the verbal descriptors affected the odor space directly or via the labels the observers might use in such an experiment. It is a significant finding in any case, because it is clear that observers do come to the laboratory with their own labeling systems for odors even though they are not required to use them in multidimensional scaling, which asks only for similarity ratings. That is, an observer may rate a pair of real odors as similar because they both smell aromatic, although he or she is not required to report that label.

Ignoring such individual differences and contextual effects, and assuming that an average odor space complete with dimensions can be determined through careful experimentation and sampling, how shall the dimensions be named? Introspective psychologists (Boring, 1942) devoted their efforts to such analysis of the mind (though not with multidimensional methods) into sensations, which they then named. In the case of odor it is an attempt to observe sensations, or the *dimensions* of the odor space derived, with the mind's eye and match each of them to a mental image of such a dimension or category. Brown (1958) points out that the question of how an image can be generic is an old, basic, and unsolved philosophical and linguistic problem. It requires that the mind's eye must know and be able to direct its attention to the essential attributes of each category, for example, *goaty*. Having defined an odor space, it is assumed that one should be able to name its dimensions; this task may be more conceptual than semantic. Naming the color blue can be conceptualized abstractly, but that is hardly what one actually does. One will very likely think of the sky or the ocean. With odors, such an approach would be even more likely; indeed it is doubtful that odors can be conceptualized abstractly at all. Odors are invariably named in terms of specific objects, such as lemons, roses, fishes, barns. One can visualize or conjure up memory images of sensations in other modalities—for example, a tune—but it is doubtful that one can do this with odors. Of course, one can visualize the yellow lemon and even make facial grimaces thinking about its sour taste, but ability to recall the pure odor experience per se is not that obvious. It is not only a matter of having a label or name for the experience of smelling a lemon. It is more conceptual, perhaps more akin to what linguists call "fuzzy concepts" (see Abrahamsen, 1972). If someone asks you to smell the content of a jar or bottle but hides the label from you, you will nevertheless be quite sure that the odor is familiar though you are not able to name it. This is called the "tip-of-the-nose" phenomenon (Lawless & Engen, 1977a), because the state one is in is similar (but only superficially so) to that of the "tip-of-the-tongue" phenomenon (Brown & McNeill, 1966). In the latter case a person can recall a letter or two, the number of syllables, or perhaps the primary stress of a word, but not the word itself. In the case of odor, the person in such a state seems to have no information about the name, but reading the definition of the source of the odor will quickly resolve the problem. By contrast, reading a definition can induce the tip-of-the-tongue state.

"In 1913 John Watson closed the bloodshot inner eye of American psychology." according to Roger Brown (1958, p. 93), and with more success directed its research to behavior and its neural correlates. In the case of odor perception one might also be more successful in pursuing a less ambitious goal (though not necessarily a behavioristic one) than attempting to describe the inner odor space for all time and all people. One suggestion along this line would be to select odorants systematically for multidimensional analysis to test the existence of specific dimensions (Engen, 1971a). For example, one might explore the effect of various stereochemical attributes on similarity judgments, and thus on the dimensionality in a specific experiment context. This was done with success in one experiment relating similarity judgments of homologous alcohols by human subjects to electrophysiological data from rats (Døving, 1966). The troublesome mental image was eliminated by refraining from asking either humans or rats to name the dimensions. There must, of course, be significant physiochemical attributes of odorants with specifiable physiological effects, and one may have more success finding the connections without the help of the mind's eye.

Absolute Threshold and Detection

After odor classification, the most common psychophysical method is to study the effectiveness of odorants in eliciting a response in a subject. It is a less ambitious approach than odor classification but also more practical, taking one odorant at a time and focusing on its intensity. The most commonly used index is the so-called absolute threshold (often abbreviated RL for the German *Reiz Limen*); however, this threshold is not literally absolute. *Stimulus threshold* is a less ambiguous description. The concept refers to the boundary on the physical concentration scale between the values a person can detect and those he or she cannot. It is usually represented by a concentration the person can detect on 50% of the trials on which it is presented. Because of its apparent simplicity, both in the methods used to define it and in the task required of the subject, the concept of stimulus threshold is the most widely used and respected (Engen, 1971b). Data on more than 100 substances have been compiled (Fazzalari, 1978; Patte, Etcheto, & Laffort, 1975). Next in line is the difference threshold or limen (DL), the index of the ability to discriminate differences between clearly detectable concentrations above the absolute threshold. Sensory psychologists and physiologists have worked on the general assumption that both RL and DL are correlated with a differential neurological response. Presumably, a minimum number of neural units must fire to produce a conscious awareness of a change in stimulation.

The problem with threshold has always been that it varies substantially both for individuals and for methods (Pangborn, Berg, Roessler, & Webb, 1964). According to a review by the National Air Pollution Control Administration (Miner, 1969), for example, the absolute threshold for hydrogen sulfide ranges from 1 to 45 mg/m^3 in air for individuals of different ages, sex, smoking histories, and places of residence, a range of about 1.7 log units! To date most of the effort in this area has been devoted to determination and selection of judges for sensory panels and the best method for measuring threshold.

Many have found the signal-detection theory to be a more fruitful approach (D. M. Green & J. A. Swets, 1966). There are two salient differences between this theory and the theory of threshold. First, contemporary detection theory questions the validity of the assumption that there is a certain cutoff point on the physical concentration scale below which there is no conscious experience of odor. Instead, signal-detection theory assumes that any concentration may be associated with conscious experience, and that the subject's criteria for what constitutes an odor interact with odor intensity to determine judgment on any one trial. Second, classical threshold theory assumes that training a subject can eliminate human error or bias from the results. In contrast, signal-detection theory assumes that such bias is inherent in this form of decision process and must be measured in each situation. Usually, it can be measured by noting the proportion of false alarms; that is, incorrect affirmative responses ("I smell it") to blanks. Whether or not detection is a continuous or discrete function of concentration is debatable, but there is convincing evidence that the likelihood of false alarm varies among observers, situations, and sense modalities in a manner predictable from knowledge of motivational aspects of the situation. Undoubtedly, the variation observed in odor thresholds may be due to such factors.

In Sweden, sensory–chemical and meteorological analyses have been used to predict how often the concentration of an emission from a certain source, such as hydrogen sulfide from a pulp mill, may exceed the expected threshold level for the odorant dispersed in the surrounding area. These predictions sometimes underestimate greatly the evidence of odor as reported by local residents who have been instructed to make observations from their residences (Lindvall, 1974). To be sure, this may be partly explained by weaknesses in the dispersion calculations, but evidence indicates that in such a situation a person is likely to err by overestimating both the incidence and duration of odor. Signal-detection theory indicates how one should correct the proportion of correct affirmative judgments, or "hits," of the presence of odor, with the proportion of false alarms. Since false alarms occur not randomly but rather in relation to the expectation, motivation, and strategy of the test subject, it is not sufficient simply to subtract them from the proportion of hits, as used to be common practice. The correction must be made according to a model that relates these two variables to each other and to concentration (D. N. Green & J. A. Swets, 1966). An index called d' is most often used. Although it first appears to be complex, the index is practical and straightforward. Also, it simplifies the selection of psychophysical methods and observes and facilitates comparison of results from different experiments. The main practical consideration is the specification of a measure of response bias. This measure is based on false alarms, and indicates how likely one is to experience odor, or to be motivated to do so, in various situations with different degrees of noise, uncertainty, and other consequences.

Signal-detection methodology has been applied to experiments in the laboratory and in the field (Semb, 1968). Results suggest that this methodology may reveal greater sensitivity than obtained with a classical method by F. N. Jones (1955) and also smaller individual differences (Corbit & Engen, 1971; Berglund, Berglund, Engen, & Lindvall, 1971). Apparent similarity of performance based on hits alone may be misleading, and one must also take into account the false-alarm rate, which may vary more than the hit rate for different individuals. In the experiment just mentioned the false-alarm rate was over 30% for one of the observers but less than 15% for the other. Such high false-alarm rates are unheard of in psychophysical experiments in vision and audition, but seem especially significant in the perception or misperception of odor. But even when such response bias is taken into account, human detection is likely to be superior to "objective" techniques of detection with so-called sensors. This holds both in general applicability and sensitivity. Moreover, these physical gadgets are inferior to people in detecting *rapid* changes in odor. Housewives without any prior experience were able to detect odors from a mineral wool plant in the field and dimethyl monosulfide in the laboratory only seconds after the odorants were presented (Berglund, Berglund, & Lindvall, 1974a, 1974b). (Recall also the ability of Mr. Weber described in Chapter 1.)

Still another study (Lindvall, 1973) was concerned with the detectability of traffic odors during the rush hours in Stockholm. Odor samples were collected on location and presented to untrained observers in an olfactometer in a mobile laboratory. These observers produced a reliable and valid index of sensitivity (d') related to physical and chemical analysis of the air samples. The sensitivity was higher on a busy city street than on a relatively pollution-free university campus nearby.

The history of failure to come up with reliable threshold values for hydrogen sulfide, for example, is probably the result of not accounting for response bias and recording only hits. Detection is a dynamic rather than static process involving psychology as well as physics and varying from situation to situation, person to person. The effect of psychological variables on sensory experience has in fact been appreciated for a long time. In 1899 Slosson reported in the *Psychological Review* a classroom demonstration on odor hallucinations:

> I had prepared a bottle filled with distilled water carefully wrapped in cotton and packed in a box. After some other experiments I stated that I wished to see how rapidly an odor would be diffused through air and requested that as soon as anyone perceived the odor he should raise his hand. I then unpacked the bottle in the front of the hall, poured the water over the cotton, holding my head away during the operation, and started a stopwatch. While awaiting results I explained that I was quite sure that no one in the audience had ever smelled the chemical compound which I had poured out, and expressed the hope that, while they might find the odor strong and peculiar, it would not be too disagreeable to anyone. In fifteen seconds most of those in the front row had raised their hands, and in forty seconds the "odor" had spread to the back of the hall, keeping a pretty regular "wave front" as it passed on. About three-fourths of the audience claimed to perceive the smell. . . . More would probably have succumbed to the suggestion, but at the end of a minute I was obliged to stop the experiment, for some of the front seats were being unpleasantly affected and were about to leave the room [p. 407].

Difference Threshold and Resolving Power

The assumption made in detection theory is that the sensory continuum is not broken up by threshold boundaries. In the difference-threshold task, one stimulus called the standard is compared with another stimulus, and the question is how likely the person is to detect the difference as the physical difference between the two stimuli is varied. Just as the absolute threshold is the value that can be detected on 50% of the trials in which a certain concentration is presented, the value of difference threshold is determined statistically as the physical difference that can be detected 50% of the time.

The research on the difference threshold for smell was started by Gamble (1898), who was followed by Zigler and Holloway (1935), Wenzel (1949), and Stone and Bosley (1965). All of these studies tested whether or not

Weber's law applies to the discrimination of perceived odor intensity. Since the value of the difference threshold (ΔI) will increase as concentration (I) increases, Weber compared ΔI as a proportion of I, $\Delta I/I$, also known as the Weber fraction. Since it turned out to be a constant proportion, and thus a generalized discrimination index, it is called Weber's law. It does in fact describe discrimination very well, especially if the denominator is adjusted to a good estimate of absolute threshold (I_0). Thus $\Delta I/(I + I_0)$ is known as the modified Weber fraction. Obviously, the detectability threshold would set the lower limit for one's ability to compare odor intensities. In olfaction the Weber fraction is about .25, based on results from several observers and odorants, but it cannot be said to be fixed or absolute. It varies with different methods and subjects (see Stone & Bosley, 1965).

The Weber fraction relates the physical value of the difference threshold to that of the standard and it is usually expressed as a percentage. Thus one can say that the concentration of an odorant must be changed (increased or decreased) by 25% before the average observer can tell that it has changed. It is this relatively high value that gives smell the reputation as one of the dullest sense modalities. For example, it may be compared with a value of .3% of the pitch of a pure tone of 200 Hz (Woodworth & Schlosberg, 1954).

This conclusion that the sense of smell is dull has been passed on from text to text. However, better stimulus control and better assessment of stimulus variability have forced a new look at this old view. A "fine-grain analysis of noise at the nose," as he describes it, was done in two experiments by Cain (1977a, 1977b). The nasal passages are, as noted earlier, complex channels. Only statistical estimates can be made of the number of molecules that reach the olfactory receptors. A variable quantity of molecules may adsorb to the walls of the nasal passages or otherwise be lost during the presentation of the odorant. Moreover, the stimulus vapor may be quite variable at the source. This problem was beyond experimental reach till the moment-to-moment variability of the vapor could be assessed with the gas chromatograph. With careful stimulus presentation, Cain's observers did very well and produced Weber fractions of less than 20% for three different odorants (n-butyl alcohol, ethyl n-butyrate, and n-amyl alcohol). This is a much better result than obtained previously. His best result was 7% for n-butyl alcohol, which is low enough to put the sense of smell, at least for that odorant, on a par with vision and audition in discriminating tones and lights presented successively. (Successive presentation is required for odorants; in vision, simultaneous presentation is possible and actually may make the comparison easier.)

The improvement in odor discrimination obtained thus far was the result of improved presentation of the odorants. If one takes into account other sources of variability, even keener discrimination is evident. Analysis with the gas chromatograph showed that the vapor delivered to the observers varied noticeably from trial to trial, with a standard deviation as high as 9.6% of the mean concentration. Taking this variability into account, according to Cain, the Weber fraction may be as low as 4.2% for butyl alcohol, again with the best results. Thus, for some odorants olfaction is as keen or keener than any other modality. In any case, one must keep in mind that the stimulus in any modality may be variable and that the assessment of such variability is the exception rather than the rule. It is a crucial consideration when evaluating the keenness or dullness of any modality, and the variability of the stimulus source and the olfactory system may be especially large.

Suprathreshold Odors: Psychophysical Scaling

It cannot be denied that the quality of odor measurement has made great strides forward. One of the most interesting aspects of this advance is that research should not and need not be limited to weak and barely discriminable odors but can deal with the real worlds of smells. Threshold methodology has been and perhaps still is a preoccupation of the odor researchers, but if the odorant is so low in concentration that it hardly has an odor, it is not likely to exhibit its characteristic quality as experienced at moderate and stronger concentrations. It is only at this level that the odor is important in pleasure and acceptance, or annoyance and rejection. Thresholds are of particular interest to those concerned with physiological problems, but they are much less relevant behaviorally.

In fact, one might say that absolute and difference thresholds represent levels at which stimulus control of the psychological response breaks down. This is why psychologists and others, as the methodology has proven its value, have devoted more attention to psychological scaling. Of course, we must not lose track of the significance of weak and subtle odors on mood and memory, which will be considered in Chapter 6.

A great deal of scaling research done during the last two decades was stimulated by S. S. Stevens' (1975) proposed psychophysical power law, which states that perceived psychological intensity grows as a power function of physical intensity. Earlier researchers had assumed either that Fechner's logarithmic law based on Weber's law applied, or that one could scale perceived magnitudes as multiples of threshold concentration (Engen, 1971c). For olfaction, there has been little, if any, empirical support for either of these psychophysical scales. The assumption that equal increments of a threshold concentration correspond to equal subjective increments does not hold for any specific compound, and, to make matters worse, may erroneously indicate different psychophysical functions for different odorants. It does not hold in the laboratory or in the application of odor scaling to air pollution (Berglund & Lindvall, 1979).

When scaling odorants with the methods recommended by Stevens (1975), the observers are asked to match perceived odor intensity against a scale of numbers or some other convenient quantitative dimensions, such as line length (Engen, 1971c). The procedure is similar to that used with rating scales that use a limited set of numbers or adjectives. The crucial difference is that a more refined measurement is involved. Observers match their perception of odor intensity on a dimension that is continuous and open at both ends. That is, they are free to select any number (or line length) in describing their judgment. The task is to select the number or line length so that it is proportional to perceived intensity. The central tendency of these numbers for a group of observers constitutes the odor-intensity scale for one odorant. Its mathematical relation to concentration defines the psychophysical function.

In agreement with research carried out on the other senses, the results support the conclusion that perceived odor intensity grows as a power function of physical intensity, concentration in this case. The following mathematical formula must generally apply:

$$R = c(I - I_0)^n,$$

where R represents perceived psychological intensity; I, physical intensity (concentration); I_0, an estimate of minimal effective value of the stimulus or stimulus threshold; c, the (arbitrary) choice of units of R and I; and n, the exponent of the function. In this case, subtracting I_0 from I straightens out the function, as it tends to become steeper for low concentrations, which are difficult to detect. (See Engen [1972b] for this and other modifications of the power function.) The parameter of greatest interest is the exponent, n, which shows how perceived intensity grows as a function of physical-intensity concentration. In olfaction this exponent might vary as much as from .07 to .7, depending on the odorant, the diluent, individual differences, and the psychophysical method (Berglund, Berglund, Ekman, & Engen, 1971; Laffort & Dravnieks, 1973). Exceptions to this rule of thumb can be found; exponents greater than 1.0 have been observed (Laing, Panhuber, & Baxter, 1978). However, a more representative value of the exponent would be around .6. Again, no exponent should be expected to be absolutely fixed and independent of the procedure used to obtain it (Berglund, Berglund, & Lindvall, 1978b). The same power function applies equally well to purer odors from laboratory chemicals and to complex odors from effluents. Knowledge of the psychophysical function can therefore be used to practical advantage in measuring the abatement of odor pollution (Berglund, Berglund, & Lindvall, 1977; Svensson & Lindvall, 1974). The fact that the exponent is less than 1 means that the rate of increase in perceived intensity does not follow that of concentration. For example, when odor concentration is doubled, odor intensity will be increased much less, about 1.5.

Because the psychophysical function obtained for 1-butanol has shown such stability in results from different laboratories and different techniques, it has been proposed that the scale obtained with it and shown in Figure 3.5 be used as a standard reference scale (Moskowitz, Dravnieks, Cain, & Turk, 1974). Note that this standard scale also includes references for absolute threshold and concentrations at which the odor becomes unpleasant (UNPL). Human observers are able to make so-called intramodal matches in olfaction; that is, they are able to match the perceived intensity of one odorant against the intensity of another odorant of different perceived intensity and quality—for example, various concentrations of butanol and hydrogen sulfide. This ability has been used to evaluate odor intensity as a function of distance from stack effluents (Grennfelt & Lindvall, 1973), to assess the effectiveness of different techniques of spreading manure in the reduction of its odor (Lindvall, Norén, & Thyselius, 1973), and to measure the intensity of poultry odors (Frijters & Beumer, 1979). Instead of numbers (or line lengths), the odors in some of these examples were matched by concentrations of hydrogen sulfide used in this case as a kind of standard scale and controlled with an olfactometer.

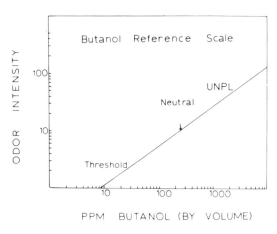

Figure 3.5. Proposed standards for expressing perceived odor intensity. The power function shown was obtained by scaling 1-butanol vapor diluted in air with the method of magnitude estimation. The equation of the function is $\log R = .66 (\log I) - .583$, or $R = .261(I)^{.66}$. Here R is subjective intensity and I is concentration in parts per million (ppm) of air by volume. Concentrations corresponding to thresholds are 2–5 ppm; neutral intensity is about 250 ppm; for higher concentrations (above 250 ppm) the odorant tends to become unpleasant (UNPL). (From Moskowitz, Dravnieks, Cain, & Turk, 1974.)

Odor-Information Transmission

For human beings, vision and audition are normally considered more important modalities for survival than taste and smell, and some would argue that perception of pain is most important. Each modality seems to do certain tasks well. The ear works well in timing events, and is also the best channel for getting someone's attention. The eye, however, is better than the ear in spatial discrimination.

The nose has been considered poor in both temporal and spatial discrimination, but as we have seen this assessment is changing. Although the difference threshold was long believed to be very large, we now learn that this is not necessarily an inherent characteristic of the sense of smell; rather, it is a result of the noisiness of the perceptual world. Added to this should be the observation by Bekesy (1964) that the sense of smell, like hearing, is keen in its ability to locate stimuli (e.g., whether an odorant is coming from the left or the right), though some claim that this ability may be largely a function of the trigeminal system (Schneider & Schmidt, 1967).

Compared with physical sensors or "artificial noses," the human nose is superior in both sensitivity and general applicability to the odors of such things as spoiled fish, air pollution, and manure. It is also difficult for artificial noses to compete with the human nose in detecting relatively rapid changes in the odorants in the atmosphere. As mentioned earlier, in a signal-detection experiment with effluents from a mineral wool plant, housewives without training detected increases in the level of dimethyl monosulfide only seconds after the changes in concentration occurred. Certain animals might be even better than humans, but the psychophysical data from animals are often of much poorer quality and not comparable (see Marshall & Moulton, 1981; McCartney, 1968; Moulton, Ashton, & Eayrs, 1960; Slotnick & Ptak, 1977). Of course, the gas chromatograph and other instruments are indispensable for the physiochemical analysis of the odorant.

The perception of odor is surely an invaluable source of information for human beings in certain situations, and probably even more so for animals in the simple psychophysical sense of providing intensity and quality information. But, like pain, odor may play a still more important part in emotion and motivation than as an intellectual channel of communication with the outside world. Odor usually involves approach and avoidance behavior once the olfactory message has been delivered. It has to do with the smell of the enemy, food, or a mate, at least in animals; we will get more and more involved in these aspects after adaptation, the reliability of olfactory information under prolonged constant stimulation, is discussed.

chapter 4
ADAPTATION

Definition of the Problem

It is a common assumption that the sense of smell, more than other modalities, is readily affected by adaptation as a result of fatigue from continued exposure to a stimulus. For example, a room one has just entered may have a noticeable odor, but presumably the odor quickly disappears because receptors fatigue and decrease their rate of firing in the presence of odorant molecules in the mucus.

Although olfactory adaptation is apparently a common experience and is generally believed to characterize odor perception, its effect has been exaggerated. It is possible, of course, that odor perception involves a short-term application for approach and avoidance of environmental stimuli, but there is surely more to it, at least for animals using olfactory cues to find food or a mate; it would be detrimental if the cue should suddenly disappear halfway there. There are in fact two other explanations that must be considered.

First, the criterion a person uses for calling something an odor sensation plays a role. A high concentration of a chemical may have a qualitatively different odor than a weaker one. Exposure to an odor may cause adaptation and thus reduce both its perceived intensity and its quality but without its disappearing altogether. Human observers can detect amyl acetate as "banana oil" at a relatively high concentration. In a weaker concentration it is not distinguishable in terms of that odor quality, but the person will still be able to discriminate amyl acetate from the "blank," although unable to define its quality (Engen, 1960; Eyferth, 1965). Such changes in criterion are an aspect of signal-detection theory, discussed in Chapter 3, and may be especially important in the case of olfaction. Although the effect of adaptation is a generally important phenomenon in sensory psychology and physiology, its intensive and qualitative time course has not been studied carefully enough to warrant any general conclusions. The main point here is that adaptation may change the quality of perceived odor and thus the criterion or the nature of the response to the odor, without causing the odor to disappear. Most experimental data deal with so-called thresholds for weak odors.

Second, sensory effects ascribed to adaptation may be due to the related phenomenon of habituation. Without specific controls, it is difficult to know whether a decrease in the stimulating effectiveness of an odorant is due to adaptation or habituation. *Adaptation* is used to describe changes in the receptors; *habituation* refers to a decrement in cessation of the response to a novel or startling stimulus. If one presents a moderately strong odorant to a sleeping infant, the infant will show a mild startle to it by moving, changing respiration rate, and making other responses. Repeating the stimulation after 30 or 60 sec will yield another response, but probably one smaller than the first. By the tenth trial the infant will most likely sleep through the whole event, oblivious to the odorant. One may erroneously conclude from this that the infant has become adapted to the odorant, but control experiments have shown otherwise. Inert mixtures of odorants (e.g., amyl acetate and heptanal) are presented until the response has reached zero. Then, on the next trial, the infant is tested with either of the two components (e.g., amyl acetate), and it is typically found that the infant responds again. It could not do this if it were adapted. The receptors still respond, so the conclusion is that the infant habituated to the mixture; the novelty wore off (Engen & Lipsitt, 1965).

The odor of a strange house or laboratory undoubtedly has a similar novelty effect. One of the primary functions of the sense of smell is to alert persons to potential hazards. But unless this alert is accompanied by some significant consequence (for example, feeling faint), they will pay no more attention to it and act as though they were not aware of it. The situation may be analogous to a ticking alarm clock in a bedroom. One is not aware of the ticking most of the time. However, when the time is important—for example, when a person is worrying about being late for work—that person will often pay attention to make certain that the clock is running. In the same way it is also possible to smell fresh paint in a room at will when one wakes up in the middle of the night, even though the odor will most likely seem weaker; and the smell of paint may not persist as well sensorily as the ticking of a clock. There may indeed be more adaptation in olfaction than in audition, but the difference has probably been exaggerated and, in addition, adaptation has been confused with habituation. Traditionally, the distinction between adaptation and habituation was made in terms of central versus peripheral receptor mechanisms. A decrease in response due to adaptation has been assumed to reflect fatigue of the receptors. A decrease in response due to habituation has been assumed to be the result of a person getting used to or ignoring stimulation no longer judged to be of significance, and thus involves central brain factors. The distinction between these concepts is becoming less clear, but adaptation is seen to involve continued exposure to stimulation that affects sensory transmission of information, and habituation involves repeated stimulation and learning. There is no physiological evidence indicating or relating olfactory adaptation to changes in peripheral receptor function. As early as 1957 Beidler proposed the hypothesis that olfactory adaptation may be mediated by a central rather than peripheral mechanism (see also Köster, 1971). Moulton (1971) likewise states that there is little evidence of adaptation in "electrophysiological recordings from the primary neurons or olfactory bulb of a variety of species: prolonged depressions of response seldom follow repeated receptor exposures to a variety of odorants [p. 69]."

The understanding of the physiology of adaptation is poor, but a number of interesting and important psychophysical studies have been made, and this chapter will emphasize that work. To begin with, there are several aspects of adaptation. What has been described so far, is often referred to as *self-adaptation*.

Self-Adaptation

Self-adaptation is the change in perceived intensity of an odorant after one has been exposed to it. In general, one distinguishes between an adapting odorant and a test odorant or target to be detected. In self-adaptation the two are the same. Usually, too, one thinks of adaptation as involving some duration of exposure to the adapting odorant, not just one sniff, and observation of prolonged exposure is referred to as a time-course experiment. The fairly fixed ideas in the literature described earlier about the susceptibility of the sense of smell to adaptation refer to this kind of experiment or real-life experience. Its history will be presented first; then we will examine the newer information showing the need for modification of the belief that now seems to prevail.

In addition to what has already been said about them, the early experiments indicate that perceived odor intensity decreases linearly as a function of the duration of the exposure to a constant adapting stimulus or source. Perhaps the best experiment of this kind was done by Woodrow and Karpman in 1917. Their olfactometer varied concentration by varying temperature of the stream of air saturated with the odorant. The concentration of molecules in a saturated airstream is proportional to the vapor pressure, which in turn depends on temperature according to Avogadro's rule. The olfactometer steadily blew the odorant into one nostril in such a way as to elicit an odor sensation without otherwise interfering with normal breathing. The odorant was presented at various levels from one to six times the concentration of an arbitrarily chosen suprathreshold level. Three well-practiced subjects were tested six times with propyl alcohol, naphthalene, and camphor. The subjects' task was to indicate by responding "now" when the odor disappeared. The average results are presented in Figure 4.1, which shows a plot of these "now" responses as a function of duration in seconds. All the functions seem to be linear, and this suggested to Woodrow and Karpman (1917) that "increasing the physical intensity of an odor by equal steps causes an increase in adaptation time by equal steps. The increase in one is directly proportional to the other. The very nature of this relationship suggests that it holds only within limits, at least that there is a lower limit, probably somewhere near sensation threshold [p. 445]." It seems logical that the stronger the odorant, the longer it would maintain sensation of odor, but it would presumably disappear eventually and sooner than one might think. The linearity of the relationship was not the most important point. Rather, it was that the odor sensation would drop to zero. Even the odor of moderately strong concentrations of camphor, the most popular substance in the early studies of adaptation, persisted only for roughly 3.5 min, and for other odorants there was no more than 5 min of function (Elsberg, 1935–1936; Mullins, 1955). It is as though a free deodorizer is built into the system; all one can

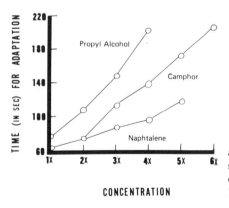

Figure 4.1. Time in seconds for disappearance of odor as a function of constant exposure. Results are shown for three odorants. (From Woodrow & Karpman, 1917.)

Self-Adaptation

or need do is to enjoy or put up with an odor, as the case may be, for a short while. That does not seem consistent with experience, at least not for unpleasant and annoying odors. The difference in slope for the three odorants would presumably indicate different stimulating characteristics. However, this method of scaling stimuli has already been criticized.

Three further qualifications must be made in interpreting these old data. First, the investigators found that the task of judging the disappearance of odor is very difficult. The observer had to learn to distinguish between sensation of odor and other sensations of taste, touch, and a trigeminal or burning sensation, in other words, what have been described as criteria here. For weak threshold sensations the qualitative differences between modalities as well as within a modality are at most faint. Some doubt that human beings can make such judgments at all (Beidler, 1965).

Second, it should be noted that Mullins (1955) and Elsberg (1935–1936) could not quite substantiate the findings of complete disappearance of odor, although Aronsohn (1886) and Vaschide (1902) did according to Köster (1971). The results have been debatable from the start, but somehow the hypothesis of complete disappearance has generally been accepted and prevails in the secondhand literature.

Third, there can be little doubt that instructing the subject to report when the odor disappears creates the expectation that it will disappear, and expectation is an acknowledged strong motivational factor. In the nomenclature of signal-detection theory, it amounts to instructions affecting the observer's criterion to be biased to "miss" the signal. In essence, the classic as well as most of the later literature on odor adaptation failed to control and measure the effect of response strategies, which are so basic to understanding perception, perhaps specially so in the case of odors with their emotional connotations.

Although this criticism also applies to the now popular magnitude estimates, Cain (1974b) has argued that this method does not suggest, at least directly, what the observer should expect. In fact, the emphasis should be, according to S. S. Stevens (1975), who promoted this methodology, that observers should be free of any constraint that might prevent them from reporting as precisely as possible the magnitude of a sensation as they experienced it.

The first olfactory experiment with such methodology was done by Ekman's group in Stockholm (Ekman, Berglund, Berglund, & Lindvall, 1967). Their observers were exposed to a constant flow of the adapting vapor of hydrogen sulfide through an exposure hood while breathing normally (see Figure 4.2). The test gas was a perceptually weak, .7 ppm of hydrogen sulfide. Judgments of the intensity of this test gas were obtained in seconds and then every minute thereafter for a 15-min session. Perceived

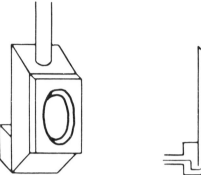

Figure 4.2. Schematic drawing of the exposure hood used in Swedish experiments. Odorous (or fresh) air flows from below and passes the subject's face at a flow rate of about .1 msec. (From Lindvall, 1966.)

intensity was judged by a finger-span technique, a variation of the direct-scaling methods (Chapter 3). Figure 4.3 presents the average results for four subjects (medical students). The initial decrease in perceived intensity is well described by an exponential function. A later experiment from the same laboratory verified this finding for other test gas concentrations (Berglund, Berglund, & Lindvall, 1977). Although there may not be enough data available to describe this function conclusively, similar functions have been obtained by Cain (1974b) for propanol, eugenol, butyl acetate, and ozone. It certainly is a better-documented description of how odor experience changes over time in this kind of situation than that proposed in Figure 4.1.

Of perhaps even greater general interest is the fact that the average curve does not reach zero. This includes Cain's results for ozone, which is suspected of anesthetizing the sense of smell. It did so only for one subject

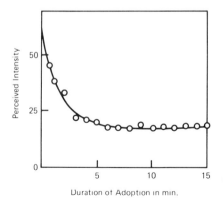

Figure 4.3. Average judgments of intensity of .7 ppm hydrogen sulfide as a function of duration of exposure. (From Ekman et al., 1967.)

Self-Adaptation

in the data shown in Figure 4.3, but then for all three adapting concentrations. But did that subject adapt, or did he have a different criterion for reporting odor? The answer to this basic psychophysical question is not known, for although the method does not constrain the subject, neither does it provide a measure of the subject's criterion for "I smell something." Generally, other experiments with different scaling methods, odorants, and subjects find that odors persist with the curve remaining above zero subjective intensity (Cain, 1974b; Steinmetz, Prior, & Stone, 1970). To obtain complete adaptation would require elimination of all sources of variability in the stimulus. That can probably not be accomplished with normal breathing, but could be in an artificial situation, for example, one analogous to that of a stabilized image in vision (Riggs, Ratliff, Cornsweet, & Cornsweet, 1953). When all eye movements are eliminated optically and the image projected on the retina truly stationary, the perceived image becomes less clear and disappears, then continues to reappear and disappear, apparently corresponding to fatigue and recovery of the receptors mediating the perceived response. The inherent variability in olfactory stimulation does affect acuity, as noted earlier, but it may also help prevent complete adaptation. The large number of receptors and other perceptual mechanisms, such as contours and contrast in the case of the stabilized visual image (Krauskopf, 1963), may also contribute to an uninterrupted function of the olfactory system by providing alternative routes for the same signal.

Another reason that the curves stay above subjective zero is that although the observers may not be told explicitly what will happen in this experiment, as in those criticized earlier, that is really what they expect because they are told to report on odor intensity. Perhaps in this case, they continue to report the experiment of odor even though it has disappeared because they believe it is still there. In the jargon of detection theory, this is a false alarm. The recovery observed when the odorant is removed and the observers are presented with fresh air instead of the adapting stimulus, interrupted with brief presentations of the .7 ppm test concentration, argues against this interpretation, that bias to perceive keeps the curve above subjective zero. However, there is no control for this in the magnitude-estimation process so that one cannot take it for granted. The instruction "tell me when the odor disappears" is surely poorer than "tell me how strong the odor is," but neither allows for measurement of how the instruction per se would affect the observer's criterion.

The Constant-Sensation Method

Instead of observing how the response changes to a constant concentration, one might observe the changes in the stimulus required to keep the sensation constant. Cain (1974b) reports the results of such an experiment for 6-min sessions in which he started the observers with one of three concentrations of 1-propanol and had them adjust the concentration dial. The results are presented in Figure 4.4 and are, as can readily be seen, very similar, possibly mirror-images of the results from the constant-stimulus method shown in Figure 4.3. Again, odor sensation does not appear to disappear but reaches a steady state above subjective zero after a few minutes. In addition, it may be significant that in both the constant-concentration and constant-sensation studies the rate of adaptation does not seem to depend on the concentration of the adapting stimulus. Regardless of concentration, the time course of the sensory event of primary interest here seems to vary in the same fashion, and relatively little in level of subjective intensity for the different concentrations. This observation probably is an indication of the arbitrary value selected by the subject in each case, but it reminds us again of the importance of such response factors in this method. The response values are not absolute but relative to each other for an experimental condition.

Effect of Intensity of the Adapting Stimulus versus Duration

There is clearly a change in perception of odor due to adaptation, and the time course of this effect seems well described by an exponential function for all concentrations. The effect of concentration as well as duration may be seen more clearly if one looks at their effects on the psychophysical power function. In early studies of this kind at Brown University

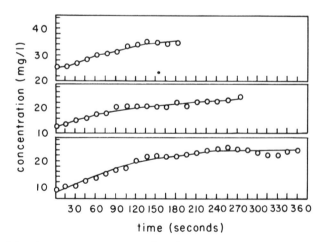

Figure 4.4. Average concentration selected to maintain constant perceived intensity as a function of duration of exposure. Results are from 9.1 (bottom curve), 12.9 (middle), and 25.4 (top) mg/1. (From Cain, 1974b.)

Self-Adaptation

(Cain, 1968; Cain & Engen, 1969), subjects were asked to sniff normally though regularly in step with a metronome, from the port of an olfactometer (Figure 3.3) delivering either different concentrations of an adapting stimulus or fresh air. Upon signal the observer would then take one sniff from another port (next to the "adapting" port) from one of several concentrations of the test odorant. These were presented in random order and judged with the method of magnitude estimation in the usual way for a psychophysical scaling experiment. Having sampled or sniffed one such odorant on a particular trial, the subject would then move right back to the port to continue the exposure from the "adapting" port. Figure 4.5 shows the effect of varying the concentration of the adapting stimulus in this situation. The results, averages from about 20 observers, show a power function with an exponent of .52 when unadapted and .30 when adapted. Although not shown here, the stronger the adaptation, the steeper the psychophysical function. In addition, a tendency of the function to become flatter for the perceptually weaker lower concentrations of the test odorant has been observed (Cain, 1970). This could be due to the tendency to sniff harder for weaker odors, thus increasing the flow rate, which in turn may increase the strength of the odorant (Tucker, 1963). However, not all have found this to be the case (Teghtsoonian, Teghtsoonian, Berglund, & Berglund, 1978), so that we cannot be sure. But, in general, the weaker the test concentration, the more it will be affected by adaptation. This is quite consistent with adaptation effects observed in other sense modalities, in particular visual brightness (J. C. Stevens & S. S. Stevens, 1963).

Once again, however, it should be noted that in this situation, too, the effect of duration of adaptation is not as expected from the old literature discussed earlier. For example Cain (1970) observed no significant effect

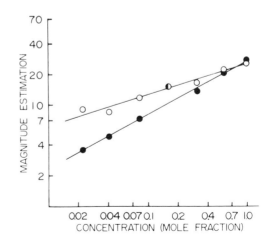

Figure 4.5. The perceived intensity of amyl acetate when exposure to each of the concentrations is preceded by exposure to the middle concentration (filled circles), and without such prior adaptation (open circles) when the middle concentration (.2) is presented only at the beginning as a reference value. (From Cain & Engen, 1969.)

from exposing his subjects to different durations as measured by the number of inhalations. In summary, (a) there is a larger effect from the intensity of adaptation than from the duration, and (b) the adaptation effect is very orderly when measured in terms of perceived suprathreshold and intensity described with a psychophysical function. I believe this is an observation of basic significance. There are other factors to be considered. Results of repeated stimulation of the trigeminal nerve leads to increase in perceptual intensity (Cain, 1976), and this might counteract a possible decrease in perceived odor intensity by stimulation of the olfactory nerve on the same trials and with the same stimulus (Chapter 2). As we shall see later, there is also evidence of facilitation with repeated stimulation in certain cases, and this could counteract adaptation.

Effect of Adaptation on Threshold

Not everybody would be impressed with data from scaling; many would prefer evidence of sensitivity from threshold data. Zwaardemaker's threshold study (1925, p. 168), presented in Figure 4.6, is probably still the most often quoted on the effect of adaptation on odor perception. He observed the effect of both duration and intensity of several different adapting stimuli in self-adaptation experiments. His unit of measurement was the *olfactie*. One olfactie is the concentration corresponding to his stimulus threshold expressed as the size of the exposed area of the tube saturated with the odorant in his two-tube olfactometer (described in Chapter 3) and 2 olfacties

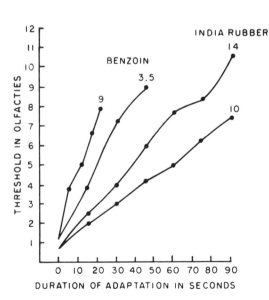

Figure 4.6. Effect of adaptation on thresholds measured in olfacties as a function of exposure to the same odorant. One olfactie is defined as the unadapted threshold value (Zwaardemaker, 1925). Results are shown for two adapting concentrations measured in olfacties (see text) for both benzoin and India rubber. (From Zwaardemaker, 1925.)

Self-Adaptation

is twice the threshold concentration. The results shown in Figure 4.6 were obtained by exposing observers for a certain period of time to a certain area while they were inhaling through the nosepiece of the olfactometer. At the end of a prescribed period of time for the experimental session, Zwaardemaker would measure threshold as quickly as possible with the same olfactometer by determining the area now required for the observers to be able to detect the presence of the odorant. Of course, sensitivity must begin to recover as soon as the adaptation is disrupted, but Zwaardemaker assumed that the recovery would be too slow to have much of an effect. This and his assumption that intensity is proportional to the exposed area in his olfactometer are both debatable.

Figure 4.6 shows nice linear effects of the duration of adaptation for 3.5 and 9 olfacties of benzoin and 10 to 14 olfacties of rubber. Briefly, the stronger the adapting stimulus and the longer the exposure, the higher the threshold (and thus the lower the sensitivity) of the subjects. These famous data also imply that sensitivity decreases in direct proportion to exposure at a fast rate, and again complete loss of sensitivity seems likely. That is not, however, what happens if one improves the methodology in certain respects.

Stuiver (1958) used a better olfactometer, an improved version of the one developed by Woodrow and Karpman (1917) described earlier. Like Zwaardemaker, but unfortunately using only himself as a subject, Stuiver made quick measurements of threshold following adaptation and found that threshold increased rapidly with duration. His threshold, in contrast to Zwaardemaker's, leveled off after exposure of longer than the 100 sec used by Zwaardemaker. The point at which perceived intensity reaches an asymptote depended on the concentration of the adapting stimulus, but it seemed to take place relatively quickly.

There is one shortcoming of all the research on the time course of adaptation, both for threshold and suprathreshold concentrations, and that is, again, the lack of a measure of the subject's response bias. This may be critical when the experimenter and the subject are the same person as in the case of Stuiver's work. Odor intensity may seem to decrease according to an exponential or linear function over time depending on the expectation of the subject. One needs an independent index of the likelihood that the subject will respond that some perceptual phenomenon has taken place in a certain context or condition independently of the stimulus that is correlated with this response. It is a matter of what decision to make, not simply whether one experiences something or nothing. One is always experiencing something. The mind of a conscious person is never empty. A payoff matrix with feedback about correct and incorrect responses will help to stabilize performance and improve the correlation between stimulus and response. That is not appropriate in the case of scaling experiments, where the human observer presumably is the only valid instrument providing readings of a suprathreshold magnitude in subjective units. But it is an intimate part of a detection experiment where the subject is simply asked to report the presence or absence of stimulation.

One experiment has been performed to determine the effect of response bias on the detection of hydrogen sulfide (Berglund, Berglund, Engen, & Lindvall, 1971) with the same equipment used in the scaling experiments (Ekman et al., 1967). The odorant was presented to the subject in an exposure hood (shown in Figure 4.2) designed to accommodate the face; the hood was edged with odorless spongy material (not shown) to provide a tight fit. The flow of the vapor is from the bottom to the top of the hood. The adapting concentration is presented in the hood on the right. The subject sat with her head in the hood breathing normally for adaptation periods of 5 min. There is a little light inside the hood, which was used to signal the subject when time was up, to exhale and then move her head to the other hood and take one sniff from it. Then she was to respond "Yes, I smell something" or "No, I smell nothing" by pressing the appropriate key on the panel just below the hoods. The test stimulus was a weak concentration near threshold for the subject. For example, it was 1.60×10^{-2} ppm for one subject as determined before the experiment. Three adapting concentrations were presented in different sessions and they were .25 log below, and .25 or .50 log units above the test concentration. (At these concentrations hydrogen sulfide does not smell too bad after a little experience.)

In order to measure response bias pure air was randomly substituted for the test odorant for half of the trials. The subject's response criterion was stabilized by paying subjects an extra two cents (actually 10 Swedish öre at the time) for each correct response (hit and correct rejection) and docking them two cents for each incorrect response (false alarm and miss). They were paid the equivalent of about $2 per hour plus the payoff earnings for participating. The results in Figure 4.7 for two of the subjects show clearly that the concentration of the adapting stimulus affects the hit rate, the proportion of correct yeses to hydrogen sulfide, and is the index of the observer's sensitivity to it. It varies in the expected fashion with the concentration of the adapting stimulus such that the highest hit rate is for the weakest adapting stimulus and the lowest is for the strongest, with the hit rate for the middle adapting concentration in the middle. All three functions are flat except for the change taking place during the very first minute. The subjects were actually tested at intervals of every 20 sec but the results are averaged to smooth the curve. The adaptation is very rapid and probably takes place immediately, but the data are too variable for a more precise

Self-Adaptation

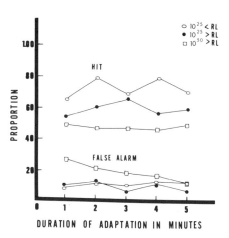

Figure 4.7. Proportion of hits and false alarms in detecting hydrogen sulfide as a function of duration of adaptation in minutes. Results are shown for three adapting concentrations measured from threshold (RL), one below and two above, obtained earlier from the subjects with the method of limits in the same apparatus. Observations were made every 20 sec, and the points plotted are averages of three consecutive observations. (From Berglund, Berglund, Engen, & Lindvall, 1971.)

mathematical description. Nevertheless, it is shown once again that even this weak odor (.7 ppm) does not disappear. Duration of adaptation is not a significant factor.

The proportion of false alarms or saying "Yes, I smell it," to pure air shows a tendency to decrease over trials. This could be the result of the subjects sharpening their criterion for giving affirmative responses and is seen most clearly for the strongest adapting stimulus and thus the most difficult judgmental task. However, the most significant observation is that the false alarm rate is very high, at least 15%, even when there is immediate feedback about the correctness of the response. The experience of odor is a difficult decision task.

The idea that odors adapt rapidly is not holding up very well. Rapid adaptation seems to take place introspectively but when tested experimentally it becomes evident that one can still smell the odor, evidence that the olfactory system continues to function just as one continues to hear even after prolonged exposure to "deafening" noise. Physiological evidence also supports this. In a study of the electro-olfactogram, Ottoson (1956, 1971) found that the electrophysiological responses from the mucosa of the frog exposed to a continuous flow of odorized air indicate that the olfactory receptors continue to respond for long periods of time. The response decreased to a lower but asymptotic level varying with concentration, but it did not disappear. He concluded that "the disappearance of the sensation of odour on continuous stimulation seems more likely to be due to the suppression of activity induced in the central olfactory pathways [cf. Adrian, 1950a] than to an inability of the receptors to respond to the stimulation [1956, p. 75]." This explanation seems similar to the suggestion made in the beginning of this chapter, that what is called adaptation may be habituation.

Recovery from Adaptation

According to Moulton (1971), "when adaptation occurs (electrophysiologically), the subsequent recovery is on the scale of seconds rather than hours [p. 69]." The psychophysical data on recovery from scaling experiments as the one shown in Figure 4.3 also indicate that recovery is as rapid as the effect of adaptation itself (Ekaman et al., 1967). The perceived intensity does not return immediately to the preadapted level but remains slightly below that for several minutes. The brief exposure of the odorant in this interval to test recovery may be sufficient to attenuate the perceived intensity. Cain (1970) also found that recovery from adaptation was quick. He specified adaptation in terms of the number of inhalations of the adapting odorant and recovery by the number of inhalations of fresh air following cessation of the adapting stimulus. He found that magnitude estimates of suprathreshold intensity showed noticeable recovery after only three inhalations of fresh air. In our detection experiment shown in Figure 4.7, there was also a test for recovery but with too scanty results to support any conclusion. However, in tests during 2 min immediately following adaptation, when fresh air was substituted for the adapting stimulus, evidence of recovery was seen for the two weaker adapting stimuli but not for the strongest. It would not be surprising to find an effect of the speed of recovery similar to the effect of adaptation itself from the concentration or strength of the adapting stimulus, such that the stronger the adaptation, the slower the recovery. In a paper on the electro-olfactogram in the frog, van Boxtel and Köster (1978) report in fact "a negative relationship between concentration of the adapting stimulus and the rate of recovery from adaptation."

Cross-Adaptation

Cross-adaptation is the name for the procedure in which one exposes a subject to one odorant and subsequently tests for sensitivity to another odorant. It is also called *coadaptation* on the assumption that different odorants will involve the same receptors and thus adapt each other. The larger the cross-adapting effect on one odorant by another, the greater the similarity of the stimulating property of the two odorants (Cheesman & Townsend, 1956; LeMagnen, 1948; Moncrieff, 1956; Pfaffman, 1951). It may be obvious why some expected that cross-adaptation results should provide a method for odor classification. Lack of adaptation should indicate that two odorants stimulate different receptors, while maximum cross-adaptation would equal self-adaptation for two odorants stimulating the same receptors and belonging to the same odor category. The method has not lived up to such expectations. Köster (1971) for example, concluded that "there exists no clear relationship between the extent of the cross-adaptation which two substances exert on each other and the similarity of their odorous qualities [p. 191]."

Cross-adaptation has been assumed to apply to even more complex and practically relevant situations of smoking and eating. Many believe that smokers would be less sensitive to odors (and tastes) than nonsmokers but this has not been demonstrated (Hubert, Fabsitz, Feinleib, & Brown, 1980). Montcrieff (1968) describes results that smoking affects the ability to detect odorants, such as pyridine, that are identical or similar to those in tobacco smoke. However, others could not find any difference in the sensitivity of smokers and nonsmokers to the odors of vanillin, 2-butanone, and ethyl alcohol, (Martin & Pangborn, 1974; Pangborn, Trabue, & Barylko-Pikielna, 1967). Similar negative findings have also been reported for taste (McBurney & Moskat, 1975). An unpublished experiment at Brown University compared the effect of tobacco and control (cabbage) cigarettes on the sensitivity to quiacol, selected because of its "smoky" odor (Engen & Furth, 1975). No difference was evident on odor sensitivity, but there was a decrease in the percentage of false alarms after smoking the tobacco, possibly because of the psychological effect of nicotine. Tobacco smoking might affect olfactory judgments, though only indirectly through some central effect rather than in a more direct peripheral fashion. Similar chronic physiological effects have been implicated in the effect of air pollution on the perception of odor (Kaiser, 1963).

The extent of cross-adaptation may still depend on one of several properties of the odorants (Engen, 1963), but one of the most common problems with the published reports on cross-adaptation has been the failure to match the odorants in perceived intensity. Matching odorants in terms of physical threshold concentrations is not adequate because it does not provide equally effective stimuli. Evidence from psychological scaling indicates clearly that stimulating effectiveness of suprathreshold concentrations varies for different odorants. To measure the extent to which one odorant cross-adapts another, one must match concentrations to be used on comparable psychophysical scales for each of the odorants (Cain & Engen, 1969). When this is done it is likely to be found that cross-adaptation has the same effect as self-adaptation with two important qualifications. Just as in the case of self-adaptation, the psychophysical function becomes steeper and the detectability of weak odorants more difficult when a subject is exposed to cross-adaptation. But the effect is smaller than in self-adapation, as Moncrieff (1956) has also observed. In addition, even when two different odorants are matched in subjective intensity, their mutual cross-adapting effects may for reasons not known be asymmetrical. For example, pentanol seems to have a strong effect on propanol, but propanol has only a relatively small effect on pentanol (Cain & Engen, 1969). This asymmetry makes the traditional hypothesis that cross-adaptation provides a simple method for measuring the extent to which two odorants excite the same receptors seem implausible. It would appear that odorants may differ in the extent to which they adapt receptors.

Facilitation

One additional unexpected finding that is detrimental to the theoretical underpinnings of using cross-adaptation to measure the similarity of odors is that in some cases the effect of odorant A may have a facilitating rather than the expected attenuating effect on odorant B. An attempt was made to describe quantitatively the time course of the cross-adaptation effect, and for self-adaptation it could be described by a mathematical function, but in the case of cross-adaptation the results turned out to be too variable to justify such analysis (Berglund, Berglund, & Lindvall, 1978a).

Facilitation was first observed in a study with newborn human infants (Engen & Bosack, 1969) in a study actually designed to compare physically similar odorants (aliphatic alcohols) through cross-adaptation. Cross-adaptation was in fact observed as expected in most cases, but in other cases the opposite happened. For example, propanol was more likely to elicit a response from an infant when it was preceded by octanol than by ethyl alcohol—one might have expected it to be less likely to elicit a response after either. Since a newborn human baby is not the usual subject in psychophysical experiments, the work was repeated with a signal-detection design with adult subjects in a more conventional way, just in case there was something special about the babies. However, the results were the same with cross-adaptation for some pairs of homologous alcohols and facilitation for other pairs (Corbit & Engen, 1971). These findings have since been supported in psychophysiological studies of detection of the same odorants by rats (Laing & Mackay-Sim, 1975). These investigators also found evidence for facilitation in the case of self-adaptation experiments, and they suggest that "facilitatory mechanisms may play an important role in chemical communication. The reception of low quantities of a pheromone or foreign odor, for example, may not result in conscious recognition of the substance, but may prime or activate specific neurons or neural circuits thereby enhancing the animal's sensitivity for the substance [p. 294]." They refer here to Nicoll (1971), who presents neurophysiological evidence from the rabbit for such a positive feedback system through monosynaptic recurrent excitation of secondary neurons.

Alliesthesia

It has been indicated that central factors may control olfactory adaptation. Homeostatic effects represent another kind of adaptive, central effect. Cabanac (1971) has suggested that the pleasantness of odors is not affected by the factors that influence their psychophysical values, but instead is determined by homeostatic factors. He describes this as *alliesthesia*. For example, it was found that although the odor of orange syrup was rated as pleasant to fasting subjects, they found it less and less pleasant following ingestion of glucose. However, regardless of their tasting condition, their ratings of the intensity of the syrup remained unchanged (Duclaux & Cabanac, 1970). Subsequent experiments (Chapter 8) have shown that not all subjects change their ratings in this fashion and the problem of different individual experiences with the test substances may be an important factor in the study of alliesthesia (Mower, Mair, & Engen, 1977).

Compensation

Still another internal factor believed to influence sensitivity is sensory compensation. It is widely believed, though probably incorrectly, that loss of any sensory function leads to improvement in some or all of the remaining modalities. Though there was actually little to support this hypothesis, it was put to the test at Brown University in a study of prelingually deaf adults (Engen & Berson, 1975). The task was to match an odor to a visually presented container; that is, when only being able to smell a bar of soap, could the person point to the actual bar of soap (Ivory, in wrapper) in an array of other visual stimuli (11 common household substances such as a can of pepper, and a bottle of vinegar). Although the hearing-impaired subjects tended to be more variable in their performance, there was no significant difference between them and the hearing subjects in this odor-recognition task. They showed no evidence of sensory compensation which was a surprise to the subjects themselves.

Although the topic of adaptation has not been exhausted yet, we are now starting to encroach on topics such as memory and hedonic judgments, which are discussed in other chapters. For example, masking was only mentioned here but will be discussed more fully in connection with mixtures (Chapter 7). Adaptation decreases the ability to perceive odors. Anosmia, our next topic, also involves the loss of ability to perceive odors, but for reasons other than overstimulation.

chapter 5
ANOSMIA

No one doubts the importance of the sense of smell in animal behavior for feeding, finding the way back to the nest, seeking a sexual partner, or avoiding the enemy. Normal development may in fact be hindered if the sense of smell is damaged (Cheal, 1975). Little attention has been paid to this problem in humans, although it is often referred to in metaphors and anecdotes. Smell is spoken of as a subtle sense, only to be invoked for hedonistic purposes according to popular conception.

It is possible for humans to be unaware of an olfactory deficit, just as it is possible to be unaware of certain deficiencies in color perception. Dalton, the famous scientist, used to startle his colleagues with his mismatched clothing; he was unaware of his sex-linked, red–green color blindness. Is odor perception, like color perception, an interesting neurophysiological problem of little significance for the survival of human beings? Schneider (1972), a physician, has stated that "physicians seldom ask their patients about the ability to smell and they rarely test this special sense [p. 272]." One may not become aware of impairment except when the loss is bilateral and total. Even then it may be thought of as more of an annoyance than a trauma, in sharp contrast to total loss of hearing or vision. However, people are now more concerned with pollution and all its possible deleterious effects. They have been alerted to hearing impairment associated with loud noises and olfactory impairment from air pollution. Strychnine and sulfuric acid may cause nasal polyps, which interfere with odor perception. There are allergic reactions to various chemical agents in the environment that involve the sense of smell. Those affected are now more likely than before to consult a physician or neurologist about such problems.

At present, the examination for odor sensitivity is likely to be very crude compared with examinations given for sight and hearing, in which orthoraters and audiometers are used. The odor test is most likely to involve the use of simple sniff bottles to determine the ability to detect and identify the odor of coffee or the like, as well as any changes in the ability to detect weak odors or to recognize different odors (Douek, 1974). There is apparently little consistency in the methods used by different physicians and little attempt to control concentration, flow rate, purity, and other potentially important characteristics of the odorants used in such clinical assessment. For example, in order to test the trigeminal response, the patient may be asked to sniff ammonia to determine if it seems to elicit a feeling, irritation, or slight tingling sensation as distinct from or in addition to odor. Such assessment is best described as a qualitative determination of the presence or absence of the ability rather than a quantitative description of the degree of acuity or loss. Psychophysicists could be of great help here, as in the other modalities (Doty, 1979).

As knowledge about the sense of smell has accumulated, it has become implicated in a wider variety of problems. For example, the absence of an olfactory response in a newborn human infant may indicate the presence of cerebral malfunction (such as forebrain fusion), chromosomal diseases, and diabetes (Sarnat, 1975). Stimulation of the sense of smell will elicit reflexes of salivation and strong breathing, which may be observed in infants within hours after birth; sensitivity increases during the first days of life (Engen & Lipsitt, 1965; Lipsitt, Engen, & Kay, 1963). There is clearly an element of startle in the baby's response, which presumably indicates the effect of olfactory stimulation on the reticular activating system. The diagnostic value of these infant responses to odorants has not yet been firmly established, but this area of investigation has promise.

Heredity

One of the principal causes of deficiencies of odor perception is heredity. Glaser (1918) was one of the first to study hereditary olfactory deficiencies; his contemporary, Blakeslee, reported in 1918 that some people perceive the odor of only some verbena flowers. For example, one person might be able to smell only pink verbenas, whereas another person might be able to smell only red ones. A study of a single family of Russians indicated that locality seemed to inbreed such deficiencies and suggested the hypothesis that the trait was sex-linked. Other investigators proposed different causes, believing that the deficiency was too common in the general population to be considered sex-linked.

Patterson and Lauder (1948) tested over 4000 people for their ability to perceive odors. They used odors selected to represent Henning's classes but obtained most of their data with a concentration of .0075% normal butyl mercaptan dissolved in 90% methyl alcohol. If their subjects could not smell it, they applied a concentration 10 times as strong. They found four types of deficiencies. The first involved four people and was characterized by a normal sense of smell except for difficulty with the weak mercaptan solution, eucalyptus, and orange oil. Although the data are too limited to conclude in favor of a single Mendelian recessive characteristic versus a more complex model, these data could so be interpreted because the group consisted of children whose parents all had a normal sense of smell.

There were only two individuals in the second group, and their problem was that they could smell the stronger mercaptan but not the weaker one. The third group involved one family in which the mother and three children, two sons and a daughter, became anosmic at middle age, although two other sons had a normal sense of smell. The authors believed that the deficiency of these children in odor perception was a delayed result of an inherited, dominant characteristic. For the fourth group the cause of the deficiency was not genetic but bodily accident and disease, to be discussed later.

• There the situation rested for a long time, during which it was also believed that albinos suffered from a general anosmia, a theory that has been disproved (Moulton & Beidler, 1967). It should be evident that the work in question leaves much to be desired in psychophysical methodology, such as the size of the groups tested and the availability of familial data. More careful studies have been made of the possible, but not yet confirmed, sex-linked recessive inability to perceive the odors of hydrogen cyanide gas. Roughly 1 in 10, and about four times as many men as women, are oblivious to the odor of such poison, but the rest of the population can detect it readily (Fukomoto, Nakajima, Vetake, Masuyama, & Yoshida, 1957; Kirk & Stenhouse, 1953). Here is apparently a case of recessive genetics, but again no definite conclusions can be drawn since most of the data have been obtained from selected individuals rather than the general population and may thus be biased (K. S. Brown, C. M. Maclean, & R. R. Robinette, 1968). K. S. Brown and R. R. Robinette (1967) concluded from such a study that the variance in the thresholds of 2885 schoolchildren to cyanide is attributable to heredity and is small compared with the total variance, which they estimated from correlations in threshold between siblings and twins of same and different sex, mother and father, and children and parents. The pattern of results does not agree with a Mendelian hypothesis, and these authors conclude that the cyanide odor threshold is not a good tool with which to study genetics since it is affected by the subjects's age and environment.

The strongest case for a genetic cause of anosmia is related to eunuchoidism, hypogonadotrophic hypogonadism, or "Kallmann's syndrome" (Kallmann, Schoenfeld, & Barrera, 1944). Kallman and his colleagues observed a sample that is as statistically representative as could be expected, of 48 such patients ranging in age from about 10 to 50 years, and their immediate blood relatives. Eunuchoid men tend to be beardless and retarded in genital development. The syndrome is much rarer in women but those afflicted have poorly developed breasts, little pubic hair, and a small pelvis. Menses are typically absent, a symptom now considered basic to the diagnosis. In Kallmann's cases, anosmia, color blindness, synkinesia (unintentional movement), and mental defects were observed regularly. Color blindness was found in both eunuchoid and noneunuchoid family members, but anosmia and synkinesia were only found in cases of eunuchoidism, according to measurement of odor sensitivity using Elsberg's technique (Chapter 3). Kallmann et al. concluded that a sex-linked genetic cause is involved in the case of color blindness but that "anosmia and synkinesia seem best accounted for as variable secondary effects of the developmental disturbance produced by the particular chromosome set-up for eunuchoidism [p. 299]." In a related study, Sparkes, Simpson, and Paulsen (1968) drew the following conclusion: "On the basis of present genetic concepts the responsible gene in the syndrome probably mediates its effect through a currently unrecognized protein, which is either biochemically important both to olfactory and certain hypothalamic functions or has resulted in structural abnormality in the hypothalamus, which affects both functions [p. 538]." It is a symptom of the neglect of this field that the study by Kallmann and colleagues, published in 1944, has only recently received much attention. It is true, of course, that people with this anomaly do not readily talk about it or submit to examination.

The anosmia associated with Kallmann's syndrome involves a not yet understood deficit in the olfactory bulb and hypothalamus, which in turn results in low gonadotropic levels and lack of stimulation of gonads. The reproductive physiology affects the perception of odor (Schneider, 1974) and patients having this syndrome also reveal "an inhibition or lack of sexual interest and arousability alone or with a partner" (Bobrow, Money, & Lewis, 1971). Kallmann et al. (1944) observed that androgonadic stimulation resulted in noticeable improvement in the development of patients' primary and secondary sex characteristics, such as increases in the size of the penis and growth of pubic hair. Later studies have reported the same finding (Males, Townsend, & Schneider, 1973; Matthews & Rundle, 1964). Still another study, of two brothers, showed similar results from such replacement theory using long-acting testosterone, 200 mg every day for 2 weeks (Sparkes et al., 1968). Unfortunately, no information is presented regarding therapeutic effects on the sense of smell, even though all agree that anosmia is an important part of the syndrome. According to Bobrow et al., "the fact that an olfactory test was not given is one of the inevitable regrets of a retrospective study [p. 333]." However, Henkin (1965) found no improvement in odor sensitivity from treatment similar to those mentioned consisting of chronic gonadotropin, testosterone, or estrogen in a study of male and female patients. However, both LeMagnen and Schneider have reported that pathologically low levels of estrogen associated with hypogonadism and ovariectomy may also be associated with poor odor sensitivity and may be helped by estrogen therapy (LeMagnen, 1950; Schneider, Costiloe, Howard, & Wolf, 1958).

Further research with larger groups is required to clarify the genetic aspects of this disease. Since it is estimated that only .2% of the U.S. population is anosmic, a thorough and systematic survey of the armed forces, or a nationwide testing program in schools comparable to that done for hearing in some states would be desirable. An attempt at such a large-scale study has been made on what Guillot (1948) originally called a partial anosmia and Amoore (1977) calls specific anosmia. Both assume a specific receptor theory, and agree with Wright (1978) that it is an inherent defect "in the sensory cells registering one of the fundamental or primary odors [Amoore, 1977, p. 267]." According to Amoore, the incidence of specific anosmia for different odorants, such as menthol, isobutyric acid, and geraniol, varies greatly among the population, depending on the sensitivity of the method and the odorant.

Amoore is systematically gathering data using a multiple-choice test involving five 125-ml Erlenmeyer flasks. Two of the flasks contain the odorant, and three contain pure water. If the observer sorts the flasks accurately, he or she is tested with a weaker concentration. The weakest concentration sorted correctly defines the person's threshold. The distribution of thresholds determined for subjects is used to set up norms to define odor sensitivity. A specific anosmia may show up as a bimodal distribution, with the anosmics showing higher thresholds than the others, as shown in Figure 5.1. But, in general, it is suggested that a person whose threshold is two standard deviations below the mean for the normal group be classified as anosmic. In a normal unimodal distribution, which apparently is obtained by Amoore as often as bimodal ones, 2.14% will be anosmic. For example, for isobutyraldehyde, the majority of the subjects have an average threshold of about 25 binary dilution steps, each step

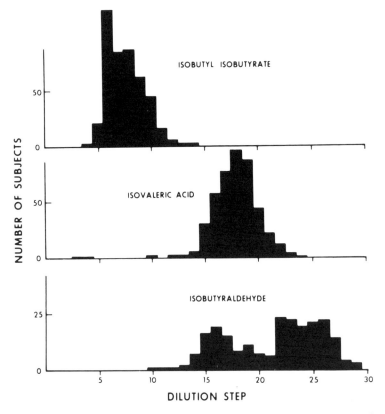

Figure 5.1. Distribution of olfactory thresholds for isobutyl isobutyrate (443 subjects), isovaleric acid (433 subjects), and isobutyraldehyde (225 subjects). The concentration is halved for each dilution step starting from 0, which represents a saturated solution. (From Amoore, 1977.)

being a concentration one-half that of the prior step, beginning with the undiluted odorant. By comparison, distribution of the minority defined as anosmic in this bimodal distribution is about 16 such dilution steps. In summary, Amoore defines anosmia not as an inability to perceive an odorant but as a high threshold. Amoore has tried to separate thresholds of panels of 30–50 normal observers from 10–30 specific anosmics on a statistical basis.

Although Amoore uses the word *anosmia,* some may prefer *hyposmia.* As mentioned earlier, Amoore is now studying anosmia for the purpose of pinning down primary odors. Subjects are tested with a series of compounds, more or less closely related chemically or odoriferously. "The size of the anosmic defect indicates how exclusively a given compound achieves the corresponding primary odor sensation. The larger the effect, the more nearly it approaches a pure primary odorant [1977, p. 270]." In addition to isobutylraldehyde, Amoore's suggested primaries include the sweaty, spermous, fishy, urinous, and musky odors represented by isovaleric acid, 1-pyrroline, trimethylamine, and 5-α-androst-16-en-3-one, and ω-pentadecalactone. For these odorants Amoore has found that 3, 16, 6, 47, and 12%, respectively, of the population have high thresholds. He believes, however, that there may be more than 30 primaries, each presumably for an odorant with a specific molecular shape that affects its own specific receptors (see Chapter 3). Pelosi and Viti (1978) present evidence for specific anosmia to the minty odor of *l*-carvone.

Although a genetic factor may be involved in certain deficiencies in odor perception, one study found no such evidence in cases of normal variation (or individual differences) in odor sensitivity (Hubert et al., 1980). The study compared the detection threshold for acetic acid, isobutyric acid, and cyclohexanone for 97 pairs of twins, 51 monozygotic and 46 dizygotic. A genetic factor would be expected to be indicated by higher correlation in the thresholds for monozygotic than dizygotic twins, but the correlations obtained for these two groups did not differ significantly.

In speculating on the coding of smell, Lettvin and Gesteland (1965) express sympathy with the idea of specific receptors but find it puzzling that the people with specific anosmia are able to discriminate other odors. This indicates to them that the condition may be more similar to a hearing loss of a certain frequency band due to injury to a limited part of the organ of Corti than to color blindness associated with the absence of a certain visual pigment. They continue:

> There is another puzzling fact. It is possible to create organic compounds that probably never occurred in nature. These will have distinctive smells, different from all others one can remember. We should think it unlikely that there are definite receptors for them. But if they affect existing receptors, why should their smells be so distinctive? From this we tend to suspect that if a receptive process could be called specific, it could only be so in the broadest sense, just as we speak of the "red" pigment that really has a wide spectrum. Counter to this view, the congenital specific anosmics ought not be so clear if the various receptive processes were very broadband and there were many different types (so that, e.g., some cyanides could affect the butyric acid receptive process, and vice versa) [p. 218].

At present, specific anosmia is defined psychophysically without any direct evidence as to how it may relate to the olfactory transduction mechanism or whether it might have genetic origin. Finding such evidence would be of fundamental value. Some hope to be able to show that specific anosmias result from inheritance or mutation, but they could also be acquired accidentally. Amoore (1971, p. 250) quotes McWhirter, who reported that about 5–8% of undergraduates at Oxford are nearly completely anosmic to the smell of freesia flowers and that "familial data show that homozygous recessives for a single autosomal Mendelian gene are anosmic." But as of now, this is only an educated guess, and firm evidence is still lacking.

The Classification of Disorders of Odor Perception

According to Hagan (1967, p. 110), the threefold classification by Specht shown in Table 5.1 has been popular among physicians for a long time. The main distinction is between *essential disturbances* (for example, damage to the olfactory nerve) and *respiratory disturbances* in the nasal passage. For example, chronic hyperplastic sinusitis can prevent the odorant from reaching the receptors. The distinction in this table between quantitative and qualitative reminds us that, on the one hand, there are degrees of loss, as in general anosmia versus specific hyposmia, and, on the other hand, there are differences in sensitivity, as in the case of normal sensitivity versus the overly keen sensitivity associated with hyperosmia.

Schechter and Henkin (1974) distinguish two forms of hyposmia. Type I often results from "surgical interruption of the olfactory nerves, extirpation of the olfactory epithelium in maxillogical procedures to treat carcinoma of the sinuses [p. 809]." The patients may still be able to smell by inhaling odorants into the oropharynx. Hyposmia may also result from laryngectomy

Table 5.1
The Specht Classification of Olfactory Disorders[a]

I. Respiratory disturbances
   A. Quantitative
      1. Hyposmia
      2. Anosmia
   B. Qualitative
      1. Cacosmia
II. Essential disturbances
   A. Quantitative
      1. Anosmia
      2. Hyposmia
      3. Hyperosmia
   B. Qualitative
      1. Parosmia
III. So-called functional disturbance
   A. Hallucinations

[a] From P. J. Hagan, Posttraumatic anosmia, *Archives of Otolaryngology*, 1967, *85*, p. 110.

(Henkin, Hoye, Ketcham, & Gould, 1968), but the reason for it is obscure, since the surgery does not involve the olfactory system. The explanation may be that there is a possible common pathway between the glossopharyngeal or vagus nerve and the olfactory nerve in the rhinencephalon and amygdala. Type II hyposmia, according to Schechter and Henkin, may be associated with loss of sensation of odor due to head trauma or hypogonadism, although the patient may not be aware of it until put to an olfactory test. The sensitivity is decreased, but not lost, because of involvement of the olfactory epithelium, and again accessory areas such as the oropharynx may also function.

These specific anosmias could also be described as quantitative, as distinguished from cacosmia and parosmia, which are qualitative. Cacosmia refers to the especially unpleasant odor of decomposing tissue associated with oral or sinus infections, and parosmia is a perversion of odor perception, which, typically involving unpleasant odors, is often associated with recovering from influenza and other viral diseases. Douek's (1974) classification is somewhat different, as is the taxonomy of the principal causes of deficiencies in odor perception presented by Schneider (1972). Table 5.2 shows that Schneider's taxonomy expands upon all the factors discussed. Organic causes of anosmia, accordingly, must be bilateral and usually involve the olfactory receptors, bulb, or tract.

Schneider states that cases involving higher levels of the olfactory system in which the patient experiences odor sensations but cannot name them are rare. Some examples of this type have been described by Mair and Engen (1976). In their test the aphasic patient is asked to point to a container identifying an odorant he is presently smelling (from an unseen widemouth sniff bottle); for example, he might be sniffing a bar of soap and being simultaneously presented visually with an array of four containers—a can of shoe polish, a bottle of whiskey, a bar of soap, and a jar of peanut butter. The more massive the lesion the more difficult the patient seems to find this task. Although that study did not relate the perceptual difficulty to specific brain functions, it seems clear that aphasic lesions affect the ability to identify odors.

The most common cause of anosmia is apparently head injury, listed in Table 5.2 under intracranial causes (see also Schechter & Henkin, 1974; Sumner, 1964). It may be that it is the most likely to be detected. Douek (1974) describes a case of complete anosmia: "She had slipped backwards on some highly polished floors striking the back of her head, but got up immediately as she did not lose consciousness. As soon as she was up she noticed that the very powerful smell of polish had disappeared [p. 137]." Any sudden acceleration or deceleration or blow to the head may cause the olfactory nerves to be sheared off at the cribriform plate through which they connect with the bulb from the epithelium. Although any blow to the

Table 5.2
Principal Causes of Anosmia[a]

Organic
  I. Intranasal
    A. Airway obstruction
      Nasal trauma
      Polyps
      Chronic ethmoid sinusitis
      Allergic rhinitis
      Carcinoma of paranasal sinuses or nasopharynx
    B. End organ disruption
      Atrophic rhinitis
      Toxic fumes (lead, etc.)
  II. Intracranial
    A. Tumors
      Meningiomas of sphenoidal ridge and olfactory
        groove or cribriform plate
      (bilateral)
      Gliomas of the frontal lobe (bilateral)
      Parasellar lesions
    B. Head trauma
      Fractures of cribriform plate
      Shearing of olfactory nerves
      Hemorrhage of the basal frontal lobes and
        bruising of the olfactory bulbs and
        tracts
    C. Infection
      Meningitis (base of frontal lobes)
      Abscess secondary to osteomyelitis (frontal or
        ethmoidal regions)
    D. Vascular
      Tumors of the floor of anterior fossa
      Atherosclerosis of anterior cerebral arteries
    E. Congenital
      Kallmann's syndrome (hypoplasia of olfactory
        system)
Psychological
  I. Hysteria
  II. Malingering

[a] From Schneider (1972, p. 273).

head can produce anosmia, the occiput seems most vulnerable. Depending on the severity of the head trauma, the result may be permanent loss of the ability to perceive odor in 5–20% of the cases. Temporary anosmia may result in others from less severe blows or from bruising of these nerves or from infections, meningitis, or other causes listed in the table. Douek (1974) notes that plastic surgery of the nose has also been known to cause anosmia, but the data regarding such incidents are poor.

The Classification of Disorders of Odor Perception

In addition to heredity and accidents, a third occurrence associated with anosmia is atrophy of receptors, which could be listed in Table 5.2 as an intranasal cause (see also Adams & Crabtree, 1961; Douek, 1974; Kaiser, 1963). There is little specific information about causative agents, but pollutants such as lead and zinc sulfate, and drugs, such as cocaine, which may upset the epithelium, are implicated. Anosmia due to damage to the epithelium and to various other intranasal causes such as influenza and environmental pollutants may only be temporary. Chances are that the patients will recover the ability to perceive odors. When the anosmia is the result of hemorrhage and edema, prospects for recovery are also good.

There is much less information provided about treatments than about causes of anosmia, but one interesting observation is that vitamin A may be helpful because it may play "an important part in the physiology of olfaction comparable to its role in vision [Duncan & Briggs, 1962, p. 116]." Presumably, vitamin A can restore the epithelial cells damaged by chemical agents or viruses. Also the prognosis for recovery from a decreased ability to smell or hyposmia is better than for total loss in cases of anosmia, which usually "remain without explanation or hope for treatment [Schneider, 1974, p. 218]."

In addition to these four general biological causes, Table 5.2 lists psychological factors affecting the perception of odors. Malingering may play a larger role than it used to in suits pertaining to annoyance with an alleged or real damage from odorous pollution (Lindvall & Radford, 1973). Some people claim to be very sensitive to and bothered by odors. They could be suffering from hysteria, but present technology usually is not capable of such differential diagnosis. Schneider (1972) describes one such case without any prior history of trauma or nasal problems:

> She was thoroughly studied including electroencephalograms of the cribriform plate and paranasal sinuses, brain scan and neurological examination, all of which were normal. Psychological testing and psychiatric observations supported the diagnosis of compensated paranoid schizophrenic illness. This patient's husband died from carcinoma of the nose eight months prior to the onset of her anosmia. This suggested that her anosmia, while not representing a conversion reaction (she gave a trigeminal response to ammonia), might conceivably represent another type of dissociative reaction such as repression of odors. This formulation is, of course, speculative, but it seems likely that her anosmia was on a psychologic basis [p. 276].

In this connection it should also be mentioned that phantasmia, olfactory hallucinations, is common in mental illness (Rubert, Hollender, & Mehrhof, 1961). Daly and Senior White (1930) present a general review of so-called "psychic reactions" to odors, and Douek (1974) notes that olfactory hallucinations are often associated with brain tumors and epilepsy.

Aging

Most assume that aging is correlated with decreased acuity in all sense modalities. Actually, the effect of aging on odor perception is not clear and has been debated since the first observations were reported around the turn of the century. Some believe that people found it more and more difficult to detect odors as they grew older, and there has been support for this down to the present (Schiffman, Moss, & Erickson, 1976). Others have not been able to verify it (Mesolela, 1943). Still others have called attention to the ever present problem of individual differences in reactions to different odors (Kimbrell & Furchgott, 1963; Venstrom & Amoore, 1968). One experiment with suprathreshold odors, different from all the others dealing with detection of weak odors, found that older subjects were as sensitive to change in concentration as the younger ones (Rovee, Cohen, & Schlapack, 1975). Health may have played a significant role, since these older subjects were predominantly healthy retired university staff. Methodological considerations are also important, especially response biases, because one's tendency to be cautious, setting a high criterion or wanting to be accurate and trustworthy, may also change with age independently of one's sensitivity. Such research remains to be done. Those who have found a decrement in odor (and taste) perception have generally assumed that the cause is atrophy and have rested their case on evidence from cadavers in which they believed they saw evidence of decline in olfactory nerves with age. There is also the belief that degeneration might be great in this modality because of deleterious effects from inhaling pollutants and because of diseases associated with the respiratory system (Smith, 1942). Both age and disease could destroy nonregenerating nerves. However, there is uncertainty about the reliability and validity of the counts of olfactory nerves. In addition, this regeneration hypothesis can no longer be assumed valid (Chapter 2). Beidler (1965) concluded that the "taste bud is seldom damaged permanently." In fact, its cells have a life of only about 250 hr, but they are continually replaced and the same applies to olfactory receptors (Graziadei & Graziadei, 1978). Beidler goes on to say that "it is useful to remember that the cells we are dealing with are not inanimate transducers but living structures that have many of the characteristics associated with other cells of the living body [p. 198]."

Some have been unwilling to conclude that age is the cause when the data suggest a decline in the ability to perceive odors. Instead, they have called attention to other factors, especially health, which may be confounded with age in such developmental studies. One example is that by Rovee et al. (1975). Another is by a group of British public health workers concerned with accidental poisoning from town gas (Chalke, Dewhurst,

& Ward, 1958). They were concerned that because of decreased sensitivity old people might be vulnerable to accidental poisoning from gas. They made a careful comparison of odor sensitivity in people over and under 65 years, half of whom were judged to be in good health and half of whom were not. The data did show that the people over 65 were on the average less sensitive to odors than those younger, but this difference was much smaller than the difference between healthy and less healthy subjects regardless of age.

Although most research has emphasized detection threshold for chemical compounds, there is currently more emphasis on the ability of the elderly to identify real foodstuffs. The elderly have been found consistently to do less well than college students in such tasks and also seem to benefit less from feedback regarding the accuracy of their performance. This suggests that the reason for the poorer performance may not be primarily sensory but cognitive; that is, a problem related to the ability to encode the perceptions of food verbally versus the ability to detect substances and perceive differences between them. Schemper, Voss, and Cain (in press) conclude that given the benefit of better feedback the elderly are able to perform better, possibly as well as college students.

It has also been suggested that another important aspect of identifying foods involves the interaction between taste and smell (Chapter 9). Murphy (1981) believes that the ability of the elderly to identify foods may be largely affected by a decline in odor sensitivity.

More research is needed on the age hypothesis (Schiffman, Orlandi, & Erickson, 1979). One very useful approach would be to test this modality as part of a general physical examination. In particular, interaction of olfactory and endocrine systems in aging is poorly understood. Current information is limited to clinical reports on the effects of menopause in women and the deterioration of the prostate gland in men. That these and other changes in the systems are related to each other and to aging seems indicated in animal research. In addition to providing important information on sensory deficits, anosmia may give useful information regarding other health problems. The effect of pollutants is but one example. However, it should not be assumed that such agents necessarily depress olfactory function. Use of cocaine first improves odor sensitivity, but it then returns to normal after about half an hour (Douek, 1974). Likewise, subjects with a blood alcohol level of 70 mg% (and mildly to moderately intoxicated) can detect odor (of quaiacol) better than control subjects (Engen et al., 1975). It has also been found that carbon monoxide (7.6% COHb from breathing air containing 800 ppm carbon monoxide) does not depress odor sensitivity by itself may do so in subjects who have been drinking and are driving a car with a leaky exhaust (Rummo & Engen, 1973). The most plausible explanation for the increased sensitivity with alcohol may be that it affects accessibility of odorant molecules at the olfactory epithelium, but it is not known how the effects of carbon monoxide, or pollutants generally, might be combined.

Hyperosmia

Hyperosmia is the opposite of hyposmia: a condition of abnormal heightened sensitivity to odor. Henkin and Bartter (1966) report that people who suffer from adrenal cortical failure (Addison's disease) are extremely sensitive to odorants. In terms of thresholds, they may be as much as 100,000 times as sensitive as observers without the disease. There is, in fact, according to the authors, no overlap in the threshold concentrations for these patients and the control subjects. Even more surprising is the finding that these patients can also smell substances such as sucrose, urea, and hydrochloric acid, which one normally will only taste, a feat approached by the normal controls only at very high concentrations. When the patients are treated with prednisolone (glucocorticoids), thresholds return to normal, and Henkin and Bartter conclude that

> these results allow the speculation that carbohydrate-active steroids, which are normally present in the nervous system, act normally as inhibitory substances in the nervous system. When the concentration of carbohydrate-active steroids in the nervous system decreases as occur in adrenalectomy, sensory sensitivity increases, i.e., the normal inhibitory action of those hormones may be removed and stimuli that normally would not elicit a response do so. When these hormones are replaced normal inhibition reoccurs and sensitivity returns to normal [p. 163].

How and where this takes place is not known, but the ability of the steroids to inhibit neural activity has been shown by others (Torda & Wolff, 1944).

On the basis of these results and the fact that sweat electrolytes in cystic fibrosis of the pancreas are similar to those in untreated adrenal insufficiency, a related study was made of a group of young people with cystic fibrosis (Henkin & Powell, 1962). Again, a dramatically heightened sensitivity in both smell and taste was observed. However, other investigators have been unable to replicate the results. One group found that such patients seem to have a normal sense of taste and smell (Wotman, Mandel, Khotim, Thompson, Kutscher, Zegarelli, & Denning, 1964). Another group of researchers could verify neither the taste nor the odor results but found the patients to have normal thresholds in both modalities (Hertz, Cain, Bartoshuk, & Dolan, 1975). Testing children does have special methodological problems (Engen, 1974). For example, some other very young autistic children are reported as having "unusual sensitivities to odorants" (Bergman & Escalona, 1949), but this refers to their preoccupation with sniffing things rather than results of psychophysical tests for sensitivity to odors.

Another observation described in terms of "relative hyperosmia" (Schneider, 1974) is the phenomenon first described by LeMagnen in 1952 that the sensitivity of women to certain musk-like and biologically significant odorants varies with the menstrual cycle. Sensitivity is high at the time of ovulation and low during menstruation. This has been interpreted as evidence of a relationship between reproduction and olfaction. Although LeMagnen limits the generalization to potentially biologically significant odorants, others claim a more general sensory superiority in women. In describing women, Rabelais (see Putnam, 1960) had this to say: "We observe in women an unusual sense of smell, and note that they flee the unpleasant and are attracted by the aromatic odors." Many other writers have made similar and more profound claims, as reviewed by Koelega and Köster (1974). These authors discuss the methodological problems confounding the evaluation of potential gender difference, for example, that gender differences in willingness to guess that an odor is present may vary as much as sensitivity with menstruation. This difference in odor perception has been linked to the hypothesis that there are human pheromones and human olfactory communication (see Chapter 8).

Effects of Anosmia

A person may not be aware of the sensory deficit and thus not suffer from it even though ability to detect certain substances quickly has survival value, making any degree of loss a handicap. However, a sudden loss of the sense of smell because of head trauma may be very disturbing, although it is often obfuscated by other primary or ancillary problems of the accident (Schechter & Henkin, 1974). Of the 29 patients described by Schechter and Henkin, the most common complaint was the impalatability of food. These authors comment that the lack of odor sensation may result because the olfactory nerves were sheared off at the cribriform plate, but this "does not explain the accompanying abnormalities of taste, since simultaneous bilateral damage to the olfactory, lingual, glossopharyngeal, and vagus nerves does not occur in these injuries [p. 809]." The problem could be that one cannot easily distinguish between experiences associated with stimulation of each of these nerves, and that what is commonly called taste is largely the result of stimulation of the olfactory nerve. Interference with odor perception may make a substance taste less strong (see Chapter 9) and affect food preferences (Doty, 1977a).

The effect of anosmia on the palatability of food is not at all trivial. What is put in the mouth must have two characteristics. In addition to fulfilling basic metabolic requirements of the body, it must also provide the central nervous system with information to regulate the intake (LeMagnen, 1971). Olfaction is not just an ancillary but a constituent part of the human neurovegetative system (Rebattu & Lafon, 1970). Nearly one-third of the patients in Schechter and Henkin's study lost weight and showed a tendency to anorexia because of the loss of pleasure in food, which in turn heavily involves perception of its odor. In another study the possibility was suggested that patients suffering from anosmia due to an injury that does not otherwise change their life-style may "eat by memory" of the odor sensations the food normally elicits (Henkin & Smith, 1971). Perhaps this is why one woman whose olfactory nerves were destroyed because of a tumor suffered no ill effects as far as her enjoyment of food was concerned (Clark & Dodge, 1955a). Or is it, as the authors of that clinical report suggest (p. 674), that "different persons vary in their dependence on olfaction in their appreciation of flavor"? (See also Clark & Dodge, 1955b.)

In still another paper Henkin and his associates postulated a syndrome of "idiopathic hypogeusia with dysgeusia, hyposmia and dysosmia" (Henkin, Schechter, Hoye, & Mattern, 1971). The syndrome characterized 35 patients suffering from taste and smell deficits who were so severely depressed psychologically that they showed signs of self-destructive impulses. But the most obvious and common symptom was weight loss—as much as 50 lb. Some of them also had vertigo, hearing loss, and hypertension associated with the onset of the taste and smell deficits.

Dysosmia or distortion of the sense of smell may be even more disturbing than hyposmia. More than half (52%) of the patients described by Schechter and Henkin (1974) as having taste and smell defects after head injury suffered from this or dysgeusia (distortion of taste): "In some patients, these symptoms were much more disturbing than the hypogeusia or hyposmia and not uncommonly obscured the actual loss of sensory acuity. Indeed, because of the intensity of these obnoxious tastes and smells some patients with dysgeusia and/or dysosmia considered their taste and smell acuity to be more acute than before their trauma [pp. 807–808]."

The psychological effect of loss of olfactory acuity is difficult to assess. One would expect that any sensory loss, even of smell, decreasing one's ability to adjust to the environment would be associated with some degree of maladjustment and occupational disablement. Loss of the sense of smell has, therefore, increasingly become involved in legal processes (Douek,

1974). Sexual drive may also be attenuated, in turn affecting reproduction. Freud and other psychoanalysts worried about this effect and they believed that normal senses are requisite for normal sexual development (Brill, 1932; Freedman, 1959; Kalogerakis, 1963; Rosenbaum, 1961). The effect of Kallmann's syndrome on young men is described by Bobrow et al. (1971) as follows:

> In 13 male patients with a diagnosis of hypogonadotropic hypogonadism, associated in five cases with verified hyposmia (Kallmann's syndrome) social development and participation were delayed. The patients associated the lack of visible signs of male puberty with their delayed sociosexual maturation. However, the evidence of physical maturation after treatment did not have the ameliorative effect on their social behavior one would have liked to predict. Dating behavior was limited before and after treatment for all the patients, and sexual interest was low even for the three married ones. The experience of falling in love was notably absent in the relationships described. Two characteristic social reaction patterns were social introversion and hostile rejection of agemates. Intellectual functioning and the incidence of personality pathology were not diagnostically noteworthy. None of the patients had any problems with gender identity except for erotic apathy which extended to masturbation. There was an absence of homosexuality. Possibly hypogonadotropic patients have a primary defect of the hypothalamus, manifesting itself as a dysfunction of the behavioral concomitants of puberty as well as a failure of hormonal puberty [p. 329].

There can also be no doubt about the importance of the sense of smell as a general arousal system, and the fact that odors easily elicit emotions and serve as excellent reminders of the past. In this way they are very important in learning and conditioning. According to Gloor (1972),

> in many animal species, motivational mechanisms involved in aggressive sexual and social behavior are activated by a rich mosaic of signals provided by the olfactory sense. This provides the organism with a highly differentiated set of cues which make possible a high complexity of behavioral patterns. These depend upon the recognition of very specific and learned, in contrast to inborn, sets of signals. The sense of smell thus becomes very important to the recall of the individual's past life experience, in the light of which current behavior can be adapted presently to existing needs. Thus, increasingly, as evolution progresses, animals are freed from the stereotyped and genetically fixed mode of operation characteristic of the hypothalamic level of organization of fundamental vertebrate drive mechanisms [p. 433].

The next chapter is devoted to the study of the associative ability of the sense of smell, or odor memory.

chapter 6

MEMORY

The popular veterinarian James Herriot (1977) provides many amusing and delightful descriptions of English farmers and their animals. Among them are some excellent examples of the associative power of odor, such as the following from World War II. Herriot had just left Yorkshire to join the RAF. The first day away was very unpleasant, involving corporals, a big city, and air force regulations added to the misery of leaving his wife, Helen. "After that first crowded day I retired to one of those green-tiled sanctuaries and lathered myself with a new bar of famous toilet soap which Helen had put in my bag. I have never been able to use that soap since. Scents are too evocative and the merest whiff jerks me back to that first night away from my wife, and to the empty feeling I had then. It was a dull, empty ache which never really went away [p. 2]."

What to most people is a "good" smell thus easily becomes unpleasant to a particular person because of a particular association. Likewise, what is to most people a "bad" smell can be quite pleasant to others. Later on, Herriot and other soldiers volunteered to help farmers with the fall harvest. One farmer grinned as he brought them to the farmyard, saying, "This is where we are going to put you through it." "But," writes Herriot, "I hardly heard him. I was looking at the scene which had been part of my life a few months ago. The cobbled yard, the rows of doors leading to the cow byre, barn, pigsties, and loose boxes. An old man was mucking out the byre and as the rich bovine smell drifted across, one of my companions wrinkled his nose, but I inhaled it like perfume [p. 216]."

Identification of Odors

To understand the perception of odors one must take into account how it is shaped through experience. It is as though the olfactory brain were a tabula rasa. As mentioned at the end of the previous chapter, the perceiver is not bound to genetically stereotyped responses but is free to develop his or her own hedonic code (Gloor, 1972).

Until the late fifties, there were only anecdotes on the subject, or what may be described as the Proustian hypothesis of odor memory. In *Swann's Way*, Proust (1928) wrote about the village of Combray, "the capital of Proust's universe," according to Lewis Galantiere's introduction to the novel (p. xiii). "Out of it come all his successful characters of his novel." The hypothesis of odor memory (actually "taste") refers to how Proust was reminded of Combray.

> Many years had elapsed during which nothing of Combray, save what was comprised in the theatre and the drama of my going to bed there, had an existence for me, when one day in winter, as I came home, my mother, seeing that I was cold, offered me some tea, a thing I did not ordinarily take. I declined at first, and then, for no particular reason, changed my mind. She sent out for one of those short, plump little cakes called "petites madeleines," which look as though they had been moulded in the fluted scallop of a pilgrim's shell. And soon, mechanically, weary after a dull day with the prospect of a depressing morrow, I raised to my lips a spoonful of the tea in which I had soaked a morsel of the cake. No sooner had the warm liquid, and the crumbs with it, touched my palate than a shudder ran through my whole body, and I stopped, intent upon the extraordinary changes that were taking place. An exquisite pleasure had invaded my senses, but individual, detached, with no suggestion of its origin. And at once the vicissitudes of life had become indifferent to me, its disasters innocuous, its brevity illusory—this new sensation having had on me the effect which love has of filling me with a precious essence; or rather this essence was not in me, it was myself. I had ceased now to feel mediocre, accidental, mortal. Whence could it have come to me, this all-powerful joy? I was conscious that it was connected with the taste of tea and cake, but that it infinitely transcended those savours, could not, indeed, be of the same nature as theirs. Whence did it come? What did it signify? How could I seize upon and define it? . . . And suddenly the memory returns. The taste was that of the little crumb of madeleine which on Sunday mornings at Combray (because on those mornings I did not go out before church-time), when I went to say good day to her in her bedroom, my aunt Léonie used to give me, dipping it first in her own cup of real or of lime-flower tea. The sight of the little madeleine had recalled nothing to my mind before I tasted it; perhaps because I had so often seen such things in the interval, without tasting them, on the trays in pastry-cooks' windows, that their image had dissociated itself from those Combray days to take its place among others more recent; perhaps because of those memories, so long abandoned and put out of mind, nothing now survived, everything was scattered; the forms of things, including that of the little scallop-shell of pastry, so richly sensual under its severe, religious folds, were either obliterated or had been so long dormant as to have lost the power of expansion

which would have allowed them to resume their place in my consciousness. But when from a long-distant past nothing subsists, after the people are dead, after the things are broken and scattered, still, alone, more fragile, but with more vitality, more unsubstantial, more persistent, more faithful, the smell and taste of things remain poised a long time, like souls, ready to remind us, waiting and hoping for their moment, amid the ruins of all the rest; and bear unfaltering, in the tiny and almost impalpable drop of their essence, the vast structure of recollection. [Pp. 54–58, from "Remembrance of Things Past" by Marcel Proust, translated by C. K. Scott Moncrieff. Copyright 1928 and renewed 1956 by The Modern Library, Inc. Reprinted by permission of Random House, Inc.]

There is another, and perhaps related, hypothesis maintained by researchers interested in olfaction: Although a person's ability to identify odors by their intensities or concentrations is very poor, the ability to recognize different odor qualities of flowers, foods, and other odorous substances is exceptional. Some have claimed that an untrained person can identify by label at least 2000 odors and an expert can identify as many as 10,000 (Wright, 1964). For a time the mathematical theory of information transmission played an important role in psychology, providing a quantitative approach to such memory and recognition problems. Human sensory ability can be viewed as a processing device for transmitting information about the environment via senses and encoding it into memory. Memory capacity, according to this theory, can be considered as the main factor limiting the ability of the perceivers. Different senses can be compared quantitatively for this factor even though they transmit information about quite different forms of stimulation such as light, sounds, and chemicals. Miller (1956) concluded his classic paper on this topic with the generalization that with unidimensional stimuli such as pitch when only the frequency of the sound is varied, the average observer is able to identify seven different levels, plus or minus two, without confusing the identity of any one stimulus (pitch) with another.

Our application of information measurement to the identification of perception of odor intensity by varying concentration showed that an average observer could identify only about three levels separated in concentration by steps exceeding the Weber fraction (Chapter 3). A highly practiced observer could reach a maximum of four, which is only approaching the low end of the range for Miller's magic number seven. Although a few papers (Desor & Beauchamp, 1974; F. N. Jones, 1968) have provided additional pertinent information, the potential effect of experience or training cannot be said to be settled. However, the results for average college students and other nonexperts are quite clear and tend to confirm the popular notion that human ability to identify odor intensity is poor. But the most interesting question is not about intensity but about the alleged enormous capacity to recognize odor quality. One must define carefully, however, what one means by identification. Those who find that olfaction is exceptional in this respect seem to be referring to the ability to discriminate between odorants presented simultaneously rather than to the ability to recognize and identify odorants presented one at a time. An average person can discriminate between thousands of different tones presented for comparison in pairs, but can manage no more than seven plus or minus two identifications for unidimensional stimuli, as we have seen. The score for tones that may vary in several attributes, such as pitch, loudness, and duration may be 100 or more but is still far from the score for discrimination. It is the same in olfaction. There are no experimental data demonstrating ability to identify such large numbers of odors. One can only find anecdotes like those described by Bedicheck (1960).

An experimental test of this problem was carried out (Engen & Pfaffmann, 1960). The observer was first presented with the odors in a group, so that he or she could compare them to clear up any possible confusions among similar odors. The observer then provided the experimenter with his or her labels whereby each odorant was to be identified subsequently. Note that these labels were the observer's own words rather than arbitrary ones provided by the experimenter, which should make the task natural and relatively easy. The observer was also allowed all the time needed for this inspection and labeling before being subjected to the second part of the experiment. The observer's task in the second part was to reproduce the correct label when each odorant was presented singly in a random order and without the help of any other sensory cues.

Our results showed that the average observer could identify nearly perfectly a set or group of odorants up to 16, but would begin to confuse them when there were more than 16 odorants in the set. Nevertheless, in terms of information theory, the amount of information remained constant at 4 bits/stimulus for these larger sets measured in $\log_2 1/p$, where p is the probability of occurrence of each stimulus in a set. The word *bit* stands for binary digit measured on a log scale to the base 2, and the antilog provides the index of the number of categories the observer could use without confusing them.

The stability of the results of 4 bits/stimulus over various experimental conditions indicates that the limitation on this score lies in the capacity of the olfactory system rather than the number of odorants presented. It is not simply that the system fails when challenged with a large number of odors, but that there is an asymptote, an upper limit to the number of odors that can be kept separately in one's mind on any one occasion. To handle more odors than that one must be able to resort to ways of organizing the odors by mnemonic devices, such as that provided by meaningfulness through personal experience. A limit of 16 is much less than expected, but it compares with performance in other sense modalities to multidimensional stimuli, such as tones varying in three or four attributes rather than just one (Miller, 1956). It is also important to note that the results were obtained with odors selected for their diversity rather than familiarity or ease of recognition because of extreme pleasantness or unpleasantness. Of course, such variables will affect the results. In one experiment we used only similar odors of acetates, aldehydes, and other sweet or fruity smelling compounds to make the task harder, and it was, performance then being about 3.8 bits/stimulus, which corresponds to about 14 categories (Engen & Pfaffmann, 1960). A similar experiment with highly distinct and familiar odors varying in pleasantness improved the score significantly, approaching 6 bits/stimulus for selected and highly practiced subjects and familiar odorants such as popcorn, beer, human urine, and cat feces (Desor & Beauchamp, 1974).

However, our own attempt to maximize performance by presenting each of 16 diverse odorants at three distinct levels of intensity, thereby allowing the observer to use both quality and intensity to identify odors, did not work. In that case the quality seemed to dominate the perception of the odor such that there was almost no information in the intensity variation (Engen & Pfaffman, 1960). We were attempting to learn what the average or typical person who happens to come upon an odor in his or her environment from an unknown source can do. For example, how well can a person identify smoke by its odor alone, (that is, in the absence of the visible evidence of fire)? The subject would do well in identifying it as smoke but not as well in characterizing the strength of the odor. Quite another problem is how well people would do if we trained them with special odors, as one might wish to do with perfumers. Cain (1979) notes that among the important factors in such training are familiarity with the substances involved, prior associations with them, and aid in recalling the label used in identifying them. Humans are excellent learners in all modalities. They can learn to identify more than 16 odor qualities, but their accomplishments in olfaction would probably take third place after those in vision and audition. Our experiments involved untrained observers and mostly unfamiliar odors, but what about the expert tested with familiar odors?

Odor Experts

F. N. Jones (1968) performed an experiment similar to ours with a "genuine chemist's nose" and two perfumers who had many years of experience in their field. Surprisingly, Jones found that these subjects performed at about the same level as nonexperts. The chemist picked out 45 odorants with which he said he was familiar, and when they were presented one by one in a random order he identified 16 or them correctly. The two perfumers selected 192 odorants, which were presented to them in sets of 16; for each of these sets they made four to five errors. One would expect them to do better, and they probably would if they were given reinforced practice in that particular task, as were the subjects in Desor and Beauchamp's (1974) experiment. Jones also considered the potential benefits of such training. However, as Turk, a noted odor expert himself, commented in the discussion following Jones' paper (p. 143), identification of single odorants is not a representative test of what a perfumer does. His task is more typically to discriminate between perfumes or perfume ingredients and make predictions about how a mixture of them will smell. He is not typically asked or trained to label odors in an absolute sense as required in these tests. In fact, we have shown in subsequent research that the use of preordained labels (especially those supplied by others) tends to diminish identification or recognition scores for odors (Engen & Ross, 1973). Of course, there are individual differences in observers, and not all odorants are equally recognizable. For example, all of our subjects recognized the unpleasant odor of pyridine with 100% reliability in all sets. And one subject (who graduated magna cum laude) always produced the highest recognition score; that is, good learners do this task well also.

The Influence of Verbal Labels on Odor Memory

The limitation of this identification method, sometimes called the *method of absolute judgment*, is that it involves a special kind of memory based on previously learned associations between odors and verbal labels. Committing information to and retrieving it from memory is not generally limited to that approach. In fact, it is a common experience that one will smell an odor, recognize that it is familiar and belongs to a general class or category, but be unable to come up with a special label for it. We have described this (Chapter 3) as the tip-of-the-nose phenomenon (Lawless & Engen, 1977) in another study comparing olfaction with other modalities, in particular the well-known tip-of-the-tongue phenomenon (Brown & McNeill, 1966).

Naming odors is a difficult task, which is undoubtedly one of the reasons for the failure to achieve a universally acceptable odor classification system. Davis (1975) suggests that it is difficult to form associations between odors and words. In our odor memory experiments, subjects seem to favor working with odor names and forming odor–word associations, but later, when the odor is presented for identification, the odor sensation does not necessarily help retrieve the word from semantic memory (see also Eich, 1978).

In one experiment (Engen & Ross, 1973), the subjects were presented with a diverse set of 20 odorants, one at a time. They were instructed to identify each odorant by label or by means of a short description of its use. The average number of correct identifications was only 6.3, or 31.5%, for a group of 40 college students. (Of course other odorants and subjects would lead to different results.) Accepting good associations as correct responses, for example describing amyl acetate as "airplane glue" or "fingernail polish" or whiskey as "booze" gives an average of 10.9 or 54.5%. Other experiments in the same study showed that the ability to recognize odors is superior to the ability to label them. Poor verbal ability in this area, an especially noteworthy characteristic of aphasics, suggests that the connections between brain centers for language and olfaction are diffuse (Mair & Engen, 1976).

In one experiment we asked subjects to match each of 20 odorants presented on cotton swabs to each of 20 labels provided on a list in an irregular order. Half of the subjects received a list of common names for the odorants (for example, *alcohol, raspberry flavor,* and *vinegar*), and the other half received typical descriptions of the same odorants (such as *antiseptic, bubble gum,* and *salad dressing*), which had been given by other subjects in previous experiments. The first list led to statistically significant higher scores (39% versus 69%) in the matching task. Moreover, when the subjects later were asked to select each of these odorants, now paired with another "new" odorant, the correct or better labels led to significantly better recognition (76% versus 70%). That is, it is difficult to come up with a proper label for an odor, but some label–odor associations are better than others (Engen & Ross, 1973).

To explore the subjects' lexicon of names or labels we asked a group of 60 to list 10 odors and for comparison 10 colors (in a counterbalanced order for different subjects) they could remember as quickly as possible (Engen & Eaton, 1975). The average scores were 48 sec and 12 sec for odors and colors, respectively, a highly significant difference. Even more interesting were the labels themselves; the most popular ones are shown in Table 6.1. In this test the individual differences were much greater for odor. For example, arbitrarily defining a label used by more than 5% of the subjects as a category leads to 41 categories for odors, but even with this low criterion almost half (48%) of the odor responses used by the subjects cannot be put into a category and must be called *miscellaneous*. For colors the *miscellaneous* label constitutes only 9.5% of the responses. *Perfume* was the most popular odor label, which was used by 35% of the subjects, *roses* was used by 23%, and *lemon* was used by 20%. By comparison *orange, yellow,* and *green* were the three top labels for color used by 90, 88.3, and 86.7% of the subjects, respectively. What is also evident is that idiosyncratic responses are typical for odors; for example, "smell

Table 6.1
Free Recall of Colors and Odors[a]

	Number of subjects		Percentage of all subjects
Labels	Male	Female	
Colors[b]			
Orange	28	26	90.0
Yellow	29	24	88.3
Green	28	24	86.7
Blue	25	26	85.0
Purple	24	26	83.3
Red	26	23	81.7
Brown	21	23	73.3
Black	19	18	61.7
Pink	13	14	45.0
White	13	12	41.7
Tan	7	5	20.0
Chartreuse	8	3	18.3
Gray	4	7	18.3
Turquoise	6	5	18.3
Gold	6	4	16.7
Maroon	1	8	15.0
Magenta	3	5	13.3
Aqua	4	3	11.7
Silver	6	1	11.7
Light blue	3	3	10.0
Light green	2	3	8.3
Indigo	2	2	6.7
Odors[c]			
Perfume	10	11	35.0
Roses	6	8	23.3
Lemon	4	8	20.0
Onions	4	7	18.3
Grass	2	9	18.3
Oranges	4	6	16.7
Bacon	4	6	16.7
Rubbing Alcohol	4	6	16.7
Flowers	4	6	16.7
Steak	3	6	15.0
Smoke	2	7	15.0

Labels	Number of subjects		Percentage of all subjects
	Male	Female	
Odors[c]			
Salt water	3	6	15.0
Exhaust	8	1	15.0
Spaghetti	3	5	13.3
Shit	7	1	13.3
Sweat	5	3	13.3
Leather	2	6	13.3
Gasoline	4	4	13.3
Ammonia	3	4	11.7
Beer	2	5	11.7
Coffee	3	4	11.7
Dog	3	4	11.7
Wood burning	1	5	10.0
Dope	5	1	10.0
Paint	2	4	10.0
Baby powder	3	3	10.0
Shampoo	1	4	8.3
Cigarette smoke	3	2	8.3
Pine	5	0	8.3
Dirty feet	3	2	8.3
Rotten eggs	2	2	6.7
Chocolate	0	4	6.7
Peanut butter	2	2	6.7
Popcorn	3	1	6.7
Wine	2	2	6.7
Garlic	2	2	6.7
Rain	2	2	6.7
Musk	1	3	6.7
New car smell	3	1	6.7
Burning rubber	2	2	6.7
Metcalf chemistry lab	2	2	6.7

[a] From Engen and Eaton (1978).
[b] These 22 labels accounted for 90.5% of the color responses. Each of the remaining 9.5% were used by fewer than 3 of the 60 subjects.
[c] These 41 labels accounted for 51.7% of the odor responses. 48.3% were used by fewer than 3 of the 60 subjects.

of a dusty old book in the John Hay library." Although such responses were obtained for colors ("color of a mallard duck"), they were unusual. It seems fitting to describe odor language as both idiosyncratic and impoverished.

Odor Memory without Verbal Labels

To avoid handicapping subjects with the verbal problem, we (Engen & Ross, 1973) adopted another method, which entails a conception of memory that is closer to the conventional one. The first part of this procedure exposes the subject to the odorant and asks him or her to pay close attention to it, rate it for familiarity and pleasantness, and try to commit it to memory. Whether or not the observer is told that there will be a later test of memory is not significant as far as the score is concerned.

The second part of this method tests the subject's ability to discriminate between these inspected odorants and others and thus show that he or she recognizes the odorants presented in the first part of the experiment. Each of the previously inspected odorants is paired with another and different odorant not previously inspected, and the task is to select the odorant that was inspected. In vision (Shepard, 1967) such pairwise discrimination between "new" and "old" pictures is excellent and far superior to olfaction, but only when the second part of the experiment follows the first fairly closely. A comparison of Shepard's visual data and our odor data is presented in Figure 6.1. Note that although the odor data are based on a

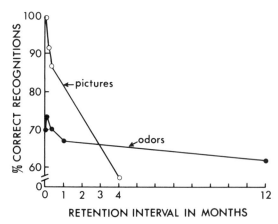

Figure 6.1. Average percentage correct recognition of the "old" stimulus in a pair of stimuli as a function of duration of the retention interval for pictures and odors. (From Engen, 1977.)

selection of an original inspection of 48 odorants, the visual data are based on over 600 pictures. Two observations are especially interesting.

As already indicated, the figure shows a striking difference between the two modalities for an immediate test of recognition within minutes after inspection. In vision the performance is nearly perfect at that point (99.7%) but in olfaction it is less than 70%, and that is for a much smaller number of alternative stimuli. But as the retention interval is increased—that is, as the second part of the experiment is delayed from 1 day to 1 month, visual recognition memory declines rapidly but odor memory remains virtually unchanged. The visual data conform closely to the classic form of the forgetting curve proposed by Ebbinghaus, but the olfactory data show a flat function that even after a year shows only a drop of 5%, which is of borderline statistical significance. And even after that long an interval after inspection, and after experience with all the odors of daily living in the meantime, the ability to select correctly the "old" member of each pair is still significantly better than the 50% chance level. Time seems to play no role in odor memory.

In psychology, whether or not there is a difference of this kind in the so-called short-term memory and long-term memory is a theoretical issue. In odor memory there is essentially no difference between them, a point well demonstrated in a short-term memory experiment similar to the one just described but with retention intervals measured in seconds (Engen, Kuisma, & Eimas, 1973). In this procedure, the subject is presented with an odorant on a cotton swab with eyes closed, and when cued by the experimenter takes one sniff and tries to remember its odor. The experimenter then announces a three-digit number chosen at random and the subject is to start counting backward by threes. The purpose of this is to prevent the subject from rehearsing or keeping his or her attention on the image or name of the odor in the retention interval. After 3, 6, 12, or 30 sec a second odorant is presented, which is either the same as the first or different, and the subject is asked whether it is the same or different. (Note that the 3-sec retention interval is too short for counting backward.) The results presented in Figure 6.2 are similar to the long-term memory data of Figure 6.1. There are no significant differences in the results for the different retention intervals, unlike retention of verbal items, which decreases rapidly when rehearsal is prevented (Peterson & Peterson, 1959).

One hypothesis is that pictures are more easily recognized than odors after a short interval because they contain many distinguishable attributes such as size and color that can be used to code them for later retrieval, and that odors represent more unitary experiences, such as those described by Herriot at the beginning of this chapter. It is just this aspect of odor perception artists refer to, this all-or-none fashion in which odors are coded.

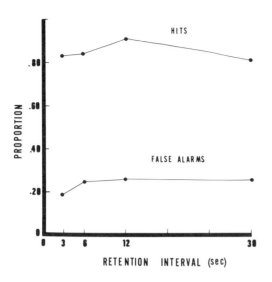

Figure 6.2. Average proportion of hits and false alarms in short-term odor recognition as a function of duration of the retention interval. (From T. Engen, J. E. Kuisma, & P. D. Eimas, Short-term memory of odors. *Journal of Experimental Psychology*, 1973, 99, Figure 1 on p. 224. Copyright 1973 by the American Psychological Association. Reprinted by permission.)

It seems to be relatively less efficient and leads to errors especially early on, but it leaves odors more resistant to confusion and forgetting later, when pictures may be confused because they share a single attribute such as color with other pictures. Of course, pictures also vary and so we are again referring to what seems *characteristic* of odor perception, not necessarily *unique*. It is, however, possible to construct unusual geometric figures that yield recognition results similar to those for odor (Lawless, 1978).

The examples of visual and odor memory are taken from the laboratory and not real life. Some of the odors or some of the pictures may have been familiar to the subject and others not, and this would influence the results. It is only the temporal difference between recognition of odors and pictures that is of interest here, not the absolute value of the scores. Not all the evidence shows the superiority of odor memory (Davis, 1977), but our data clearly support the observations of so many writers, as well as individual experiences, that the memory of odors is a special, if not unique, psychological phenomenon. The flatness of the odor function in Figure 6.1 is a robust phenomenon. However, certain factors will affect it.

Factors That Influence Odor Memory

In addition to training and the nature of the label, other factors affect odor memory. One such factor is the number of alternative odorants inspected originally. In one experiment, for example, 22 odorants were inspected rather than 48 as used for the data in Figure 6.1, with the result that the whole function is moved up significantly to about the 85% level on the ordinate but stays fairly parallel with the present results (Lawless & Cain, 1975). Likewise, in the short-term memory experiment (Figure 6.2), requiring the subject to inspect and remember five odors rather than one significantly decreased the number of hits and increased the number of false alarms (Engen, et al., 1973).

Another factor affecting recognition performance significantly is, not surprisingly, the similarity of the "old" and "new" member of the pair, as in the recognition of spices such as basil, bay, celery, marjoram, and oregano in a short-term recognition experiment (Davis, 1975; Jones, Roberts, & Holman, 1978; Mair et al., 1980). For example, subjects are more likely to recognize the odor of onion as having been inspected earlier if it is paired with Scotch whisky rather than garlic. We have observed this effect especially clearly in Korsakoff patients who suffer olfactory deficits (Potter & Butters, 1980). Although they show evidence of rapid forgetting in a short-term memory task in audition and vision when intervals between inspection and recognition test vary between 5 and 30 sec, their ability to recognize odors does not show this decay over such short intervals. However, if the odors are too similar, their performance approaches chance for all retention intervals (Mair et al., 1980). In other words, the Korsakoff patients are impaired in perception of similarity but not in memory for odors. This suggests another hypothesis in addition to the unitary perceptual odor image, namely, that visual memory and olfactory memory involve different brain structures. The neurological impulses in the olfactory system seem to have a more direct route from the receptors to the brain. They have direct access to the limbic system and then to the cerebral hemispheres. Olfactory information may therefore be processed more quickly and with less editing than visual and auditory information. Odor memory may last longer because of a larger number of connections to different parts of the brain that may make possible more associations (Engen, 1980).

Regarding the notion that perception of odor is closely tied to emotion, neurological evidence indicates that the right temporal lobe is more important to short-term recognition memory of odors than is the left lobe (Rausch, Serafetinides, & Crandall, 1977). Patients who had undergone anterior temporal lobectomy to relieve epilepsy were tested in a short-term odor recognition experiment similar to the one mentioned earlier (Figure 6.2), with several trials with four alternative odorants (amyl acetate, phenylethyl alcohol, pyridine, and cyclopentanone). The mean scores for groups of about 10 subjects each were 53.6, 80.4, and 96.3% correct for right temporal lobe excision, left temporal lobe excision, and normal volunteers without any history of neurological dysfunction, respectively. It is interesting that, compared with the left temporal lobe, the right is more involved in nonverbal functions. The results of this study also seem consistent with the present notion regarding the role of verbal labels in odor perception.

Forgetting

Forgetting has generally been believed to involve primarily inhibition or interference of related experiences taking place between the original presentation of the stimulus and the test of memory of it later on. Assume that one of the odors inspected was garlic. This is a substance that one might encounter subsequently, for example, at an interesting dinner party before the test of recognition memory, and one would therefore have another association with garlic that would compete with the experimental one. The present results indicate that odor memory is not affected greatly by such intervening experiences, or retroactive interference, because there is little effect from the length of the retention interval with increased likelihood of forming such other associations.

Odor memory is, however, influenced by proactive interferences, in which forming one association with a stimulus may make it more difficult to acquire others subsequently. This problem has been explored formally in so-called paired-associate learning experiments (Davis, 1975; Lawless & Engen, 1977a). In our experiment the subjects were asked to associate each of 12 odors with very distinct pictures, scenes from different foreign countries taken from travel magazines. The odors and pictures were presented twice in two different pairings. Subsequently, the subject was asked to point to the picture now presented in an array of all the pictures that had been previously paired with that odor. The results show that the first of the two pairings or associations was far better retained than the second. In other words, it is difficult to overcome the impact of the first association; it interferes with the memory of the second one.

That learned associations to food odor and taste have survival value is dramatically illustrated by "bait shyness," an animal's avoidance of food that has made it sick (Garcia & Brett, 1977). Many people have experienced the same kind of lifelong aversion to a particular food or drink after overindulging or consuming it coincidentally with the onset of illness. Unpleasant odors have been used successfully in aversion therapy to treat overeating by presenting them while the subject was eating a favorite food (Foreyt & Kennedy, 1971). Children receiving chemotherapy for cancer with drugs that have gastrointestinal side effects of nausea have acquired an aversion to a novel ice cream, "mapletoff," which was tested for that purpose as part of their diet during this treatment (Bernstein, 1978). The general problem is that such treatment-induced discomfort may lead to loss of appetite and, in extreme cases, anorexia nervosa. Subsequent pleasant experiences with the same food do not readily change such aversions, and just the odor alone may recapture the original feeling. In other words, proactive inhibition may prevent the unlearning of the unpleasant association and the replacing of it with a pleasant one. In the case of food aversions established experimentally in rats, it has been demonstrated that the electrophysiological response to food odor in the olfactory bulb is markedly different after a food aversion has been established (Pager & Royet, 1976). The reason that odor is such an effective stimulus in food aversions, according to one hypothesis, may be that it acquires some of the properties of taste, which in turn facilitates its access to central mechanisms of taste where it may be categorized as a food stimulus and be protected from interference by other odors, thus leaving olfaction free to monitor the environment (Palmerino, Rusiniak, & Garcia, 1980).

The difference in how retroactive and proactive interference seem to affect the more cognitive senses of sight and hearing and the more affective senses of smell and taste is also consistent with the many anecdotes about odor memory. Sometimes, it seems, the problem is that we cannot forget about experiences elicited by an odor, odor memory being too good, as Boris Bedney (1962) describes in his short story "Mosquitos":

> The aromatic resinous smoke unexpectedly reminded Voskoboynikov of the half-forgotten scent of incense; his mother had been religious and had taken him to church in his childhood. He thought how unfortunate it was that his memory could retain this ancient smell, this early rubbish, to the end of his days, while it would forget many more recent and more important things. This subconscious contraband was like dirt tracked in from the outside; he'd always carry it with him even to communism itself, these old and unnecessary memories [p. 60].

Is There Odor Recall?

Although odor memory is fairly impervious to the effect of time, it is not necessarily outstanding in terms of the number of items that can be committed to memory, and it would be overstating the case to conclude that odor memory is unique. The level of training and the sophistication of the odor are important considerations, and so is the nature of the task defining how memory is measured. It seems to be a rather special ability pertaining to recognition, or identification of a present stimulus, rather than absolute recall of a stimulus from previous experience. A quote from Nabokov (1970) will state this point better: "She used a cheap, sweet perfume called 'Tagore'. Ganin now tried to recapture that scent again mixed with the fresh smells of the autumnal park, but, as we know, memory can restore to life everything except smells, although nothing revives the past so completely as a smell that was once associated with it [p. 60]."

A few people claim they can upon command recall specific odors. However, it often seems that such memories result from the confusion of odors with other sensations. As was mentioned in Chapter 3, one can remember the color and shape of a lemon and even make grimaces associated with the memory of its sour taste, but without being able to conjure up the odor experience per se. Unfortunately, there seems to be no simple experimental method for testing the existence of such recall (see Ellis, 1928, p. 56).

chapter

7

ODOR MIXTURES

In real life one is rarely exposed to pure tones or monochromatic lights. One is even less likely to smell the odor of a pure compound, and such odorants are difficult to obtain even for research purposes. Most of the odorants encountered are mixtures. Our concern here is with so-called inert mixtures. What is smelled is usually a psychological mixture resulting from the activities of the olfactory system in response to two or more odorants presented simultaneously, not the result of one chemical interacting with another.

The study of the perception of odor mixtures is a problem of genuine interest, on a par with perception of mixtures of colors or sounds associated with their respective physical stimulus correlates. It is not an exaggeration to say that the perception of odor mixtures is of greater interest for applied psychology than perception of pure odorants. Mixing odorants is an essential aspect of the perfumer's work, and must also be of interest to the chef. It should also be of great concern to industries, which in concert emit different odorants into the atmosphere. Air pollution may be caused by only one odorant, but like noise pollution it is probably the result of several odorants being present simultaneously, or of poor ventilation (Cain, 1979a). Theoretically, such a situation may result in a stronger odor because the components add psychologically and one odorant may even intensify (or potentiate) another. However, the suppression of perceived odor intensity is also possible because there may be antagonism between the components (Baker, 1964).

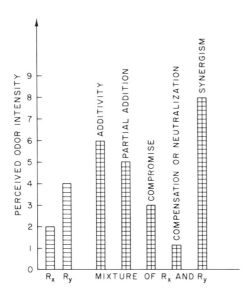

Figure 7.1. Possible results (R_{xy}) from mixtures of the perceived intensities of two odorants R_x and R_y. (Based on Cain & Drexler, 1974.)

Olfactory processing of mixtures of odors involves the perception of both intensity and quality, and we shall deal with both. In the case of pollution or any other form of overstimulation of the olfactory system, one typically attempts to reduce intensity by manipulating the source and/or to manipulate the quality with a deodorizer, which usually means adding another pleasant odorant. The intensity problem is easier to deal with both conceptually and experimentally, and has by far received the most attention. Figure 7.1 depicts the possible effects on perception when two odorants are mixed, including the classic problem of compensation (Chapter 1).

Perceived Intensity of Odor Mixtures

When mixing two odorants, both above detection threshold, one typically finds that the perceived intensity of the mixture is less than the arithmetic sum of the individual intensities but more than their average (F. N. Jones & M. H. Woskow, 1964). For example, if a magnitude-estimation experiment indicates that the subjective value of odorant A is 20 and that of odorant B is 10, then the perceived intensity of the mixture of the two will be less than 30 but more than 15. Cain and Drexler (1974) evaluated the masking effectiveness of a mixture of different amounts of substances used as deodorizers—linalyl acetate, linalool, lavandin, and a mixture of lavandin and linalool. They scaled different concentrations of these "deodorizers" mixed with pyridine, an unpleasant odor, and found that the mixtures were generally weaker than the sum predicted from the individual components of the pyridine and deodorizer smelled separately. This finding is consistent with the knowledge that the psychophysical function shows perceived intensity to be a negatively accelerated function of concentration. Each added step on the concentration scale of the odorant becomes perceptually smaller and smaller (Chapter 3). Like most sensory modalities, the sense of smell tends to compress physical input.

In addition, it may be characteristic of odor perception that it does not discriminate well between changes in quality versus changes in intensity, at least partly because the two are correlated (Engen, 1964). Even though the experimenter is changing only the concentration of an odorant, this action may change its quality as well, and quality may dominate in perception.

Adding odorant A to odorant B might primarily affect perceived odor quality rather than intensity, but the research on this topic has almost invariably concerned perception and measurement of odor intensity, which appears to present a simpler psychophysical problem. One may perhaps observe a clear-cut intensity effect in the case of odors that are weak, presumably below the "detection threshold." The perceptual sum obtained by adding a so-called subthreshold concentration of one odorant to the suprathreshold concentration of another makes the latter odorant smell stronger (Baker, 1964; Rosen, Peter, & Middleton, 1962). This effect has been described as synergism and has obvious implications for the perceptual aspect of odor pollution and annoyance. Synergism in odor perception is a rare phenomenon and probably only takes place with weak stimuli. It may be related to the phenomenon described earlier as facilitation (Berglund, Berglund, & Lindvall, 1978a; Corbit & Engen, 1971; Engen & Bosack, 1969; Laing & Mackay-Sim, 1975). The way two components combine varies with their intensities, and Köster (1969) suggests that synergism is rare at suprathreshold levels and then is usually observed when the amount of the two components of the mixture is even. Köster and McLeod (1975) obtained similar results from electrophysiological recordings from the olfactory bulb of the rabbit and psychophysical responses from human observers, suggesting that the perceived odor intensity of mixtures depends on peripheral factors such as the EOG discussed earlier.

Cain (1975) performed several magnitude-estimation experiments with mixtures of suprathreshold odorants and in all cases found that the perceived intensity of the mixtures of 1-propanol and amyl butyrate was less than the sum predicted from their intensities when judged separately. In addition, he also showed that this result holds not only for the typical case of physical mixtures, either in the vapor phase or in liquid form inhaled through the same nostril, but also for so-called dichorinic mixtures when odorant A is presented to one nostril and odorant B to the other. Yet the intensity of these dichorinic mixtures is closer to the arithmetic sum of the components judged separately, falling between the perceived intensity of propanol and amyl butyrate. This result has been described as a *compromise* to distinguish it from compensation and counteraction when the sum is perceived as weaker than the stronger component and stronger than the weaker component (Figure 7.1). Dichorinic mixtures rule out competition among molecules for receptor sites and suggest that summation takes place at a level higher than that of the receptors. Neurologically, there is bilateral interaction of input from the two nostrils and therefore opportunity for such summation (Cain, 1977a). I shall return to that connection later, but here the point is that the most typical results involving natural sniffing show only partial addition. That is what one could call the first of two principles of odor mixing. It should be noted that compensation, the bone of contention between Zwaardemaker and Henning (Chapter 1) has been resolved largely in Henning's favor.

Perceived Quality of Odor Mixtures

In mixtures of odors, the perceptual quality of one odorant may dominate another. Most experiments like those described have studied intensity and its reduction or counteraction. When the problem is odor quality, and especially its unpleasantness, one speaks of *masking*. Consideration of odor-quality changes has primarily come as a result of the correlation existing between quality and intensity. Strong odors are generally disliked and are usually considered "pungent," not just strong. Deodorizers may have a qualitative effect just because they mix with the malodor. The mixture of the smells may be less intense and thus less unpleasant than the malodor, but presumably also less pleasant than the odor of the typical deodorizer. When a deodorizer like lavandin is added to a malodor such as the solvent pyridine, the intensity and the unpleasantness of the malodor are reduced (Cain & Drexler, 1974). However, such generalization cannot be made without qualification because the effect depends on the concentrations of the odorants. If the concentration of the malodor is increased its perceived intensity may grow faster than that of the lavandin deodorizer added to counteract it. In fact, under some circumstances adding more deodorizer to the environment may increase the overall odor intensity and thus be counterproductive. Strong odors are generally displeasing, and deodorizers may only be effective with relatively weak odors. It should be noted that such effectiveness may be purely psychological and aesthetic. Harmful odorants should, of course, be removed rather than masked.

In one experiment (Engen, 1964) an attempt was made to have the mixtures smell equally strong and also to match the intensities of the components so that, overall, neither one of them would dominate the other. The mixtures consisted of 100–0, 87.5–12.5, 75–25, and 0–100% amyl acetate and heptanal, respectively, emphasizing amyl acetate to counteract the more dominating heptanal. The mixtures were strong but well matched in intensity, and the observers were also instructed to ignore such intensity differences and concentrate on quality.

These mixtures were presented in all possible pairs (15) to groups of observers for individual judgment with the method of ratio estimation (Engen, 1971c). In a random order, each of the stimuli served as the standard and the other member of the pair served as the comparison stimulus. The standard was called 100 and the task was to assign the comparison stimulus a number reflecting the percentage of the quality of the standard that was perceived in the comparison stimulus. Other scales were used also, and in general it was an easy task even for untrained observers (university students) to judge similarity in quality. The average judgments are summarized in a matrix and evaluated in terms of the concentration of either odorant in the mixture (Figure 7.2). The similarity of the heptanal–amyl acetate mixtures is evidently a simple monotonic function of the relative amount of each odorant, analogous to the way in which perceived intensity is a monotonic function of the concentration of one odorant in an odorless diluent. The fact that the curves fitted by eye do not go through the origin

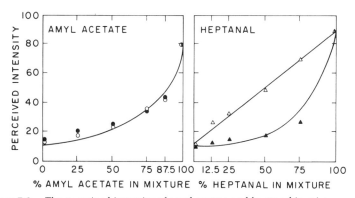

Figure 7.2. The perceived intensity of amyl acetate and heptanal in mixtures of amyl acetate and heptanal. Open circles represent unadapted judgments and filled circles represent judgments following prior adaptation to undiluted amyl acetate and heptanal. (From Engen, 1964.)

shows that there is a small degree of similarity between heptanal and amyl acetate. Most interesting is that different functions are obtained for the two components. The function for heptanal is a linear function of the heptanal concentration of the mixtures, whereas the function corresponding to the amount of amyl acetate is first lower and flatter but accelerates rapidly toward 100% concentration. The quality of heptanal evidently dominates but does not completely counteract the contribution of amyl acetate.

The Nature of Masking

On the basis of research in other modalities, such as vision and taste (e.g., McBurney, 1969), adaptation would be expected to throw light on the nature of the interaction between the two odorants in the mixtures just mentioned. Such adaptation experiments showed that heptanal had a much greater effect on the judgment than did amyl acetate. In these experiments the observer had to take a strong sniff of either 100% amyl acetate or 100% heptanal before judging the pairs. Figure 7.2 shows that heptanal adaptation depresses the heptanal function so that it becomes similar in form to the unadapted amyl acetate function. By contrast, adaptation with amyl acetate has no appreciable effect on the amyl acetate function, at least for this adapting concentration. Such asymmetric adaptation effects have been observed before (Cain & Engen, 1969) and could be an important aspect of deodorizing, especially if the underlying stimulus factor could be revealed. The unadapted amyl acetate function already looks like an adapted and positively accelerated function. Study of individual pairs shows that the effect of adaptation, as expected, is to make the observer less sensitive to the same stimulus component in the mixture. As a result, judgments of similarity, which were also obtained in the study of the two members of a pair, may either increase or decrease for the same reason. For example, prior exposure to 100% heptanal will increase the perceptual similarity of a mixture of amyl acetate and heptanal versus 100% amyl acetate, because the heptanal contribution of the mixture will be depressed. However, if the same mixture was paired with 100% heptanal, the prior heptanal adaptation would decrease the similarity of the pair because the adaptation depresses the heptanal component in the mixture. Closer inspection of the present data also indicates that the effect of adaptation is greatest for stimuli in the middle of the range, where the unique quality of each of the two components is least noticeable. This is what one would expect from the general observation that adaptation affects the perception of weaker stimuli more than strong ones. This is also why counteractants

The Nature of Masking

will not be effective against malodors of high concentrations (Cain & Drexler, 1974).

These results are consistent with reports that the quality of an odor will change with continuous exposure to it. For example, Nagel (1897, as reported by Pfaffmann, 1951) found that when the two were mixed, vanillin masked coumarin, and that adaptation to vanillin unmasked the coumarin component of the mixture. We should be able to unmask the amyl acetate function by adaptation with heptanal because previous exposure to heptanal reduces the perceptual intensity of the heptanal component in the mixture. But, by the same token, there is also a tendency for the amyl acetate component to be increased as a function of heptanal adaptation. In other words, there is evidence of both adaptation to one stimulus and enhancement of the other from adaptation to one of them. Thus, the curve shown for amyl acetate in Figure 7.2 may be the prototype of an adapted or masked function, and the curve shown for the unadapted heptanal may be typical for the unadapted or unmasked function.

Masking is obviously related to adaptation and is difficult to distinguish from cross-adaptation when prior exposure to odorant A decreases the intensity of odorant B. One might therefore expect to have observed an effect of cross-adaptation in the present case with a decrement of the amyl acetate compound with heptanal adaptation and vice versa, but if there was such an effect it was too small to discern relative to the adaptation effect observed.

The difference observed in adaptation for the mixtures of two odorants may provide clues about the physical attributes desired in a mask. For example, one idea is that the molecular weight and volatility of the odorant molecules might affect accesibility at the olfactory receptors, according to a gas chromatographic model (cf. Mair et al., 1978). Another hypothesis involves the "complexity" or oiliness of the odorants (cf. Cain, 1975). Amyl acetate may be less "complex" than heptanal. Cain and Drexler's (1974) experiment provides data relevant to this hypothesis (the more complex the odor of the masking agent the more effective it is). When both linalool and linalyl were added to the malodor pyridine, the masking effect was greater than when only one of them was added, in terms of comparable preceived intensities. The authors speculate that varying odor complexity is effective either because it is analogous to varying the bandwidth of a sound to mask auditory noise frequencies or because a complex mixture has a quality that is simply more difficult to describe and characterize. However, it must be borne in mind that the effect was still inversely proportional to concentration, so that for a high concentration of malodor complexity was not a factor.

Where in the olfactory system do the interactions take place? It may be at the level of the mucosa, at the receptors or bulb, or in the central nervous system, where it could involve other psychological mechanisms affecting olfactory sensory information processing. Cain's (1975) dichorinic data favor the hypothesis that the interaction takes place above the receptor level; yet little is known about absorption of molecules into this mucus at the receptor level and what events may affect the rate of exchange of molecules there. In fact, none of these different possibilities can be rejected, but that is a topic beyond the present knowledge of olfactory physiology.

Psychophysical Theory

Relatively few theoretical papers have been published on the quantitative analysis of the perception of odor mixtures (Lawless, 1977; Moskowitz & Barbe, 1977; Spence & Guilford, 1933). Not much progress was made when the emphasis was on physical solutions to the problem rather than on psychological solutions. Most of the interest is presently centered around the vector model (Berglund et al., 1976; Berglund, Berglund, Lindvall, & Svensson, 1973). Like most of the studies in this field, the vector model deals with intensity—data are still needed on simultaneous variation of both intensity and quality of components of mixtures. The basic idea of the vector model is that the perceived intensity of an odor mixture is the vector sum of the components of the mixture. The model depicting this is illustrated in Figure 7.3, which shows the plot of one component against another when they are smelled separately. R_x and R_y represent the perceived

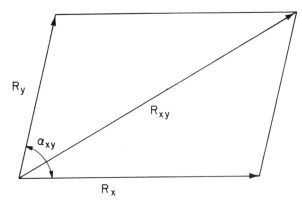

Figure 7.3. The vector model depicting the theoretical interaction of the perceived intensities of two odorants (R_x and R_y) in a mixture of odorants X and Y. (From Berglund, Berglund, Lindvall, & Svensson, 1973.)

intensity or quantity of each component. The length of the vector, R_{xy}, represents the intensity of the mixture of the two odorants. The angle, α_{xy}, is assumed to give a measure of the qualitative similarity between the two components (e.g., amyl acetate and heptanal), and it is assumed to remain constant for different mixtures of different concentrations of the two components. When α is zero the intensity of the mixture of the two components is described by the sum of R_x and R_y, although it is not a likely result, as noted earlier, except possibly for highly similar components. As the component becomes perceptually more different in quality, α would increase. At 180° the qualitative difference between R_x and R_y would be the absolute difference between their values.

The test of the model consists of comparing the judged intensity of the mixture with that of the component in a ratio-scaling experiment. In the following simple case, the intensities of the two components have been matched so that $R_y = R_x$ in intensity at various levels, and then intensity judgments of mixtures of R_x and R_y have been obtained. One next plots the value of the intensity of the latter mixtures against the sum of the individual values of the two. The plot can be used to compare different models of how the intensity will sum. Of the most interest would be, of course, an arithmetic sum. An example using mixtures of several components is shown in Figure 7.4 (Berglund et al., 1976). All the obtained intensity values of the mixtures are again below the values predicted from either the geometric or the arithmetic mean of the two components judged separately. However, the vector sum seems to fit best; it is obtained by fitting a function to the data plotted and using the slope of the linear fit

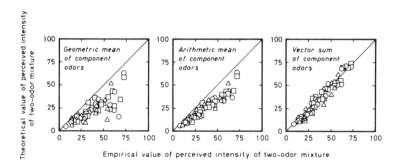

Figure 7.4. Test of three models of odor mixtures with ethyl acetate and pyridine (circles), ethyl acrylate and acetone (triangles), and pyridine and acetone (squares). (From B. Berglund, U. Berglund, & T. Lindvall, Psychological processing of odor mixtures. *Psychological Review*, 1976, *83*, Figure 1, p. 436. Copyright 1976 by the American Psychological Association. Reprinted by permission of the publisher and the author.)

as the estimate of α and then computing the vector sum of the formula for the case in which R_x is equal to R_y in intensity (Berglund, et al., 1976):

$$R_{xy} = (R_x + R_y) \cos 1/2\alpha_{xy}$$

When the components are equally strong the intensity of the mixture is simply linearly related to the intensity of any of the components, with the slope reflecting the number of odorants mixed and the nature of their interaction.

This model has been extended and tested, with promising results, for mixtures of two to five components, and it has also been applied to odorous pollution associated with effluents from pulp mills involving various sulfides and methyl mercaptans (Berglund, Berglund, & Lindvall, 1973, 1977; Greenfelt & Lindvall, 1973; Lindvall, 1970). These results show that substances that are themselves nonodorous contribute to the perceived intensity of the complex odorous gases of pulp mills. Alternative models on the same theme have been suggested (Patte & Laffort, 1979), but there are not enough data available to favor one model over the other. However, Moskowitz (1979b) argues that although the model presented here works for binary mixtures a new model is needed for mixtures of three or more components based on the angular separation of vectors obtained from binary mixture experiments.

According to Cain (1975), the value of α obtained in several experiments has fallen within a narrow range of 105–130°. This suggests that different odor mixtures are processed in a similar fashion and could conceivably be related to the power function for the intensity of the single odorants and thus predictable from the ratios of concentrations. Both the power function and the vector model describe perception as an attenuating process. In general, the model suggests that odorants mutually inhibit each other in such a fashion that the intensity of a mixture of odors exceeds the intensity of one of the odors perceived singly by a relatively small amount. It should be noted also that this is a model that depends on psychological values for practical predictions, such as the degree to which a single pollutant contributes to the intensity of perceived pollution.

Although the model has been tested in the case of odor intensity, it might be expected that the perceived quality of a mixture would be most similar to the strongest component of the mixture. This result was in fact obtained in the experiment described earlier (Ekman, Engen, Künnapas, & Lindman, 1964; Engen, 1964) in which the quality of mixtures of amyl acetate and heptanal was scaled. The same is indicated by the research of Cain (Cain, 1975; Cain & Drexler, 1974). All results point to the difficulty of masking strong malodors with weak masking agents. Future research must emphasize quality and show better ways to deal psychologically with strong malodors.

Is the Sense of Smell Analytic?

The vector model assumes that the mixture of odorants will produce a homogeneous percept, one that is experienced as an entity. However, that does not mean "nonanalyzable" (Shepard, 1964) or "integral" (Garner, 1970). Such extremes can perhaps be found in odor perception; for example, two different sulfides, though distinct substances, might have such similar odors that they could not be separated in a mixture experiment. Such a case might be analogous to attempting to see the separate wavelengths of a light comprising a color. Vision is often classified as a synthetic modality for this reason; it is not considered possible to sort out the components perceptually (F. N. Jones & M. H. Woskow, 1964). The sense of smell is usually thought to be analytic, like hearing, in which different frequencies can be discriminated as qualitatively different pitches. However, one wonders if the whole distinction has any practical value. For example, even in vision observers certainly can judge an orange light in terms of its yellowness or redness, which may be a good example of a homogeneous analyzable mixture.

Some time ago the distinction was made between metathetic and prothetic dimensions (cf. S. S. Stevens, 1975). The distinction is one between quality and quantity, or kind and magnitude as far as psychological dimensions are concerned. Loudness is prothetic and pitch is metathetic. A change in pitch is not a change in intensity but in quality, a change in kind. In terms of the scaling criteria used to make the distinction (Eisler, 1963; Engen & Lindström, 1963), judgment of odor intensity was correlated with changes in odorant concentration, yielding scales that conform to the prothetic scale, but barely. Perceptually, both quality and intensity often change with concentration, and one cannot reject the hypothesis that psychophysically the sense of smell is a metathetic dimension. Perhaps the best way to characterize it is to say that it primarily transmits information of qualitative similarity of percepts rather than psychophysical differences in intensity. The fact that quality is an important characteristic in perception of odors can hardly be doubted—as the next chapter will try to show.

chapter 8
ODOR HEDONICS

The perception of odor serves first of all to arouse attention and to orient the person in identifying the source of the odor and judging its significance. It is not clear how well human beings can localize odors, but the ability is well documented in animals. Research has produced the concept of pheromones or secretions emanating from one member of a species to affect other members (Beauchamp et al., 1976). Although there are disagreements about the definition of a pheromone, there is no doubt that such odorants can initiate sexual activity, serve as warning signals, and delineate trails and territory. Examples of such behavior are too numerous to list here (see Dröscher, 1969); but bombykol, released by the female silk moth (*Bombyx mori*), will drive a male within several miles upwind to the female. The provocative thought is that such chemical messengers may exist in humans; it has probably been the single most important factor in renewing interest in the sense of smell, illustrated by the emphasis on sexually arousing perfumes.

There is another reason for this renewed interest, namely, the possibility that the sense of smell may be useful in warning people of contaminated food and air. Perception of odors may actually dominate visual perception in the selection of food items (Beauchamp & Maller, 1977). The food may look fresh, but the smell will make all the difference, if the food is actually spoiled. Moreover, we use this modality for monitoring the air we breathe.

In humans, as in animals, olfaction may encompass endocrine effects with far-reaching behavioral results, including the selling price of real estate (see Lindvall & Radford, 1973). Surveys in the United States indicate that about one third of the complaints received by air pollution authorities from citizens were concerned with unpleasant odors, often in the absence of air pollution violation by industry (Shigeta, 1971). Malodorous contaminants in the home associated with insulation and aerosols have also become a concern. Even deodorizers used as masking agents have been criticized because they add to the unwanted materials in the atmosphere. However, such situations present complex problems of measurement. Indoor air quality can be specified by the detectability and intensity of odors, and thus psychophysics can provide quality criteria used to evaluate methods of ventilation (Berglund & Lindvall, 1979; Cain, 1979a).

Problems in Hedonic Measurement

The measurement of the pleasures or annoyances associated with odors has generally been considered to be less precise than psychophysical measures of intensity, which have received most of the attention from researchers. Zwaardemaker (1961) commented that when a naive subject is asked to judge the intensity of an odor, he or she may be unable to follow the experimenter's instructions, and that people are "generally not accustomed to distinguishing accurately the qualities and intensities of smells they meet. . . . Some discover chemical resemblances; others are strongly impressed by the agreeable or disagreeable effects connected with the sensations [p. 502]." A generally accepted conclusion from multidimensional scaling is that the pleasantness and unpleasantness of odors account for most of the judgments of the similarity of odors, but that this hedonic dimension actually is quite complex psychologically and entails more than simple likes versus dislikes (Moskowitz, 1979a; Schiffman, 1979; Woskow, 1968).

In lecturing about the "pleasures of sensation," Pfaffmann (1960) noted the apparent characteristic differences in sense modalities. On the one hand, sight and hearing are keen senses, as indicated by difference thresholds. Very small changes in a stimulus will be detected by the typical human observer. On the other hand, however, sights and sounds have relatively little effect on motivation and emotion, compared with the effect of the sense of smell. Hedonics actually seems to characterize the perception of odors, and unpleasant odors apparently play a greater role than pleasant ones. For this reason unpleasant odors have been applied in the treatment of both alcoholism and overeating by aversion therapy (Foreyt & Kennedy, 1971; Kennedy & Foreyt, 1968) as an alternative to stimuli such as electric shock. In such therapy the alcohol or dessert may be associated with unpleasant odors such as skunk oil and butyric acid through classical conditioning, and may in fact come to elicit the displeasure associated with the unpleasant door.

One survey (Hamauzu, 1969) found only 20% of the estimated 400,000 odorous compounds to be pleasant; the rest were unpleasant or neutral. In a study of a relatively small sample (110) of odorous chemicals and household items, the typical individual judged about half of them to be unfamiliar or strange (Engen & Ross, 1973). In turn, only 11% of these unfamiliar odors were rated as pleasant, whereas 50% and 39% were judged as unpleasant and neutral, respectively. There were great individual differences in preferences for specific odors, but there did seem to be a strong tendency to judge an unfamiliar odor as unpleasant. One individual would judge an odor familiar and pleasant, whereas another person would judge the same odor unfamiliar and unpleasant. The suspicion of the unfamiliar is evident in the use of the sense of smell; it alerts and warns the perceiver, putting him or her in a state of arousal. It is to be expected that the brain structures associated with odor perception are in the same region as those mediating unpleasant emotions.

Individual Differences in Odor Preferences

Individual differences are inevitable. In the case of unpleasant odors from a sulfate mill, such differences have been observed for gender, health, and age. According to the participants at the Fourth Karolinska Symposium (Lindvall & Radford, 1973),

> the classes of variables which are most relevant to annoyance surveys include: Level of awareness of sources of environmental pollutants, feelings or affective responses to these sources, duration or periodicity of the reaction, salience of the response, and demographic, sociological, and economic characteristics. Other related variables are information level and feelings about environmental problems in general, social awareness of annoyance issues, and attitudes toward the source of pollution, such as beliefs about its potential of harmful effects [p. 9].

Despite the problem of subjectivity of data obtained from human observers, they are the target population and cannot be ignored. A survey of public opinion about diesel exhaust indicates that odors might be rated more objectionable than they actually are because of prejudice regarding pollution in general (see Springer, 1974). There is no simple psychophysical relation between exhaust, for example, and odor, and irrelevant factors can bias the judgments. The color of smoke may bias an observer to report that it smells (Engen, 1972a). Petitioners in a public health case reported that they were much more bothered by offensive odors than were the "silent" majority of the same population (Cederlöf, Friberg, Jonsson, & Lindvall, 1964). On the other hand, those making their living at a factory are less likely than others to report annoyance with its odorous effluents and may in fact find the odor pleasant. Psychological factors then may play a decisive role. In a study (Cederlöf, Jonsson, & Sörenson, 1967) of aviation-caused noise pollution in a city in Sweden, a group of 270 respondents was divided into two halves. Half the group received a book presenting a favorable view of Swedish military aviation and similar interesting information, whereas no special attention was paid to the other half. When an opinion survey on air traffic was conducted a month later, 18% of the first half compared with 43% of the second half reported being disturbed by airplane noises.

In another study (Cederlöf, Friberg, Jonsson, Kaij, & Lindvall, 1964) a result on a personality questionnaire indicated a relationship between personality traits and adverse effects of odors. Annoyance with odors was often found to be combined with reports of nausea and headaches, leading the authors to conclude that "The results also show that the annoyance is due not only to the exposure question, but also to factors among those exposed. Thus, it is clear that annoyance is more frequent among those reporting previous respiratory or cardiovascular diseases and also among persons with a propensity to neurosis, sensitivity to other external environmental factors and propensity to displeasure with other aspects of the community [p. 44]."

In addition to the problem of response bias, the use of opinion polling to measure the intensity of experience may provide misleading data in another way. It is tempting to interpret the frequency of a certain response in an opinion survey as an index of how great the annoyance is or how strong the odor must be, but one must distinguish between the existence of an effect and its magnitude. The same percentage of people in two different cases may report displeasure with a malodor, one response having been elicited by a strong odor, the other by a weak one. For example, the odor of diesel exhaust is generally found to be both stronger and, probably more disagreeable than gasoline exhaust (Degobert, 1977). There is need for research bridging this gap between psychophysical scaling of intensity and opinion polling regarding annoyance (Berglund, Berglund, Jonsson, & Lindvall, 1975). For example, one study (Engen & McBurney, 1963) suggests that quantitatively the hedonic aspect may represent a significantly greater cause of stimulation than perceived intensity. Although the so-called dynamic range (the ratio of the weakest to the strongest stimulus) for perceived intensity was only about 10:1, the range obtained was as much as 150:1 for judgments of the least pleasant/most pleasant odor. Although these ratios depend on the odorants sampled, they do provide a good indication of an important difference.

Allocentric versus Autocentric Responses

Another methodological problem is that the mode of response may differ according to sense modality. Visual experience seems to be more likely to involve thought and cognition, and olfaction to stimulate emotional or motivational arousal. The visual and auditory systems seem inherently capable of abstracting information such that one's sights and sounds are not experienced in the raw but are spontaneously organized in a cognitive and intellectual way. Little or no such information is involved in olfactory experience. Schachtel (1959) believes that visual and auditory sensations are analyzed in an object-centered or allocentric mode and that olfactory sensations are evaluated in a subject-centered or autocentric mode. The distinction between these two modes of perception lies primarily in the degree to which the stimulus is perceived as existing independently of the perceiver. In olfaction, perception is characterized by how the object feels to the perceiver, not by what the object is like. The degree is "pleasure-boundedness," that is, the degree of sensory pleasure/displeasure or comfort/discomfort associated with the perceptual experience, is another way to distinguish between these modes. This difference is reflected in the way people respond to stimulation. Responses to odors are relatively emotional. Since olfaction apparently is closely related to brain centers involved in arousal and emotion, but has weaker links with the centers of language, it may be more difficult to verbalize a response to an odor than to a picture or a sound. Retrieval of appropriate words would therefore be predicted to be relatively more difficult for odors than for pictures (Lawless & Engen, 1977b). Evidence consistent with this was obtained in an experiment in which subjects were required to free-associate to pictures or odors representing the same six common objects (Engen & Omark, 1977). Subjects were instructed to tell what either picture or odor reminded them of. Associations to odors tended to involve less explicit and less accurate noncontent vocalizations, such as "Umm's" and "Ah's" with longer pauses, less fluent statements with smiles and grimaces and comments about likes and dislikes. For example, the odor of onions led to this response: "Yuck [pause]—umm, oh, wait a minute [long laugh]—this is going to sound stupid—between roasted lamb and burning rubber," and the odor of Lysol led to this one: "This reminds me of a hospital—the first one—ah, but [pause] it reminds me of a hospital, let's say a maternity ward or where they keep the kids—it's sweet smelling. Well, I like that, it smells good." In contrast, pictures of the same stimuli elicited more content words and more stimulus-oriented commentary such as "Uh, that reminds me of, uh, working in a pantry [pause] at my food services job" in response to a picture of onions, and "Sterile hospital alcohol, ether" to a picture of a Lysol can.

Origin of Odor Preferences

Perceptions as distinguished from raw sensations of odors may be considered psychological in that they are acquired or modified by experience in the environment. Pheromones may be special in that they represent stimuli having innate functional properties, but it is difficult to find other examples. For example, dog repellents would be of great interest to many homeowners, but so far no such odorant has been found. Repellents, food flavors, and perfumes serve the intended purpose generally as the result of marketing, advertising, and conditioning. Odors per se do not contribute any specific behavior or performance. For example, perfume does not cause any specific mood; it only initiates a response in the olfactory system. Odors may thus be considered as nonfunctional components of objects, but they may, however, become powerful stimuli through association with gratification of hunger, or sex, or achievement of success (M. A. Mitchell, K. S. Konigbacher, & W. M. Edman, 1964). Probably, what is innate is the propensity to learn to make such associations, and the close relationship between the olfactory system and the parts of the brain mediating emotional responses, not the hedonic values of the odors themselves. Odors that are not associated with significant events will, through habituation, be ignored. This is true to some extent of all modalities. Habituation can take place even with stimuli too weak or obscure to be clearly recognized (Kunst-Wilson & Zajonc, 1980).

There is evidence from experiments with newborn rats and other animals that early experience with odors, even aversive ones (Cheal, 1975; Cornwell, 1975; Kaplan, Cubicciotti, & Redican, 1979), in the home cage leads to preference for these odors. Studies of children suggest that except for the restive novelty response just referred to, odor preferences are absent at birth and acquired with age. An early study seemed to show agreement in the preference ratings of odors by children aged 7–9 and adults, but the authors did express concern about the individual differences in reliability of the ratings (Kneip, Morgan, & Young, 1931). Had they observed younger children, greater differences would probably have been found. Although individual young children may have preferences (Steiner, 1977), the younger the child, the less reliable is the prediction that can be made of that child's preference on the basis of the preference of adults (Engen,

1974). Moncrieff (1966) said that children seem more tolerant of odors than adults are. A psychoanalytic study of the development of odor preference implying innate preferences is usually mentioned (Stein, Ottenberg, & Roulet, 1958). On the basis of Freudian theory of psychosexual development, the authors believed that there was a discontinuity in the hedonic ratings of body odors plotted as a function of age (Figure 8.1). Up to age 4 the children said "Yes" more than "No" to the question of whether they liked the odors of (synthetic) sweat and feces, whereas children 5 years and older responded "No."

We repeated the study (Engen, 1974) and obtained similar results (Figure 8.2) but discovered that the results for children 4 years and younger were totally biased because they answered "Yes" regardless of whether or not the question was phrased in the negative or in the positive; that is, both "do you like it?' or "do you dislike it?" yielded affirmative and thus contradictory answers. The break in the preference curve with age is the result of problems of communication rather than odor.

There is little if any firm evidence that newborn babies, 2–4 days old, show any preference, measured in terms of their facial impressions, stabilimetric activity, and breathing pattern for such odors as asafoetida or anise (Engen & Lipsitt, 1965; Engen, Lipsitt, & Kaye, 1963; Self, Horowitz, & Paden, 1972). The largest response in our study was to acetic acid, a trigeminal stimulus, then asafoetida, phenylethyl alcohol, and anise. All the facial and bodily responses to the odors, whether pleasant or unpleasant to adults, looked like mild startle and escape reactions. Steiner (1977)

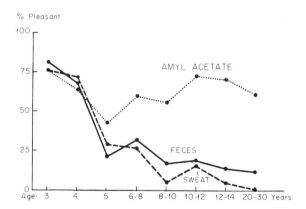

Figure 8.1. Percentage of individuals in each age-group (with 20–60 members) who responded that they liked the odor of amyl acetate, feces, and sweat. (From M. Stein, D. Ottenberg, & N. Roulet, A study of the development of olfactory preferences, *American Medical Association Archives of Neurological Psychiatry*, 1958, *80*, 264–266. Copyright 1958, American Medical Association.)

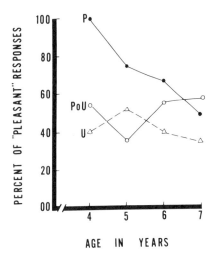

Figure 8.2. Percentage of responses indicating that an odor was pleasant as a function of age (with 17 to 44 members in each age-group) combined for odorants ranging from pleasant to unpleasant (safrol, amyl acetate, diethyl phthalate, and butyric acid) in order to assess the effect of the form of the question. P and U show results from "Does it smell pleasant?" and "Does it smell unpleasant?" with a choice of affirmative or negative answer. PoU represents "Does it smell pleasant or unpleasant?" with choice of responding "pleasant" or "unpleasant." (From Engen, 1974.)

photographed the reactions of newborns to different tastes and smells. Sweet tastes seemed to elicit facial expressions suggesting satisfaction and bitter tastes disgust and apparent dislike, according to observers judging the photographs. Likewise, the odor of vanilla, for example, seemed to elicit pleasure and the "fishy" odor of artificial shrimp flavor seemed aversive.

Another experiment showed that 2-day-old human infants tend to turn away from ammonia, another trigeminal stimulus (Rieser, Yonas, & Wikner, 1976), but they might also turn away from a pure odor that does not stimulate the trigeminal nerve, especially if it is strong. Such control experiments separating odor intensity and quality and stimulation of first (olfactory) versus fifth (trigeminal) nerves cannot be done with intact human subjects. Adults whose input from the trigeminal nerve has been obstructed tend to judge odors as being weaker than observers with normal trigeminal function (Chapter 9).

Experiments have shown that, presumably because of its odor, infants turn more often toward a breast pad that had been worn by their own mother and was presented to one side than toward one from another breast-feeding mother that was presented on the other side (Macfarlane, 1975). At 2 days of age the child turns to either pad more or less at random, but at 6, and then at 8–10 days, the average percentage of times the infants turn toward their own mothers increases from 57.8 to 60.3 and 68.2%, respectively. Russel (1976) also showed that infants are more likely to make orienting responses and begin sucking toward their own mother's breast pad rather than to that of another mother. Again, this response was not

Origin of Odor Preferences

present at 2 days of age but was at 2 weeks and became even stronger after 6 weeks. More evidence has been obtained that odor perception plays a role in a child's attachment to its mother at 3–5 years of age (Schaal, Montagner, Hartling, Bolzoni, Moyse, & Quichon, 1980). Mothers can also identify their baby's smell, according to Russel and Mendelson (personal communication). They asked blindfolded mothers to smell babies, all wrapped in identical blankets with only the eyes exposed. Each mother was asked if she could tell whether a baby was her own. Mothers were able to do this, not at all perfectly, but at a substantially significant level even after only a few minutes of actual exposure to their children after birth.

Even children of 1 or 2 years of age show no strong evidence of differential response to unfamiliar odors (L. P. Lipsitt, et al., 1977). The test situation, shown in Figure 8.3, consisted of a table with several interesting toys in front of a pictorial display shielding an experimenter and equipment for presentation of odorants. Once the child had become engaged in playing with the toys, we attempted to determine whether presentation of an odorant would interrupt this play and whether the child's reaction to the odorant could be judged as pleasant, neutral, or unpleasant by observers behind a one-way mirror in the next room. The mother also rated the child's reactions. The odorants were presented with atomizers through inconspic-

Figure 8.3. Odor preference test of a 2-year-old at the Behavioral Development Laboratory of the Child Study Center at Brown University.

uous holes in the pictorial display. Using water as a control, the tests consisted of several trials each with two pleasant (amyl acetate and lavender) and two unpleasant (dimethyl disulfide and butyric acid) odorants. The test odorants were all quite strong and had been equated for perceived intensity by adults. Each odorant was presented twice and in an irregular order with ample time between odorants to air out the room. The results with 50 one- and two-year-old children showed the neutral response to be the most typical for each age-group. Although the weak ordering of the odorants was similar to that for adults even when eliminating the data from the children (about 40%) who failed to indicate any hedonic response of like or dislike whatever on any of the 13 trials, analysis of variance provided no evidence of differences in the ratings of the hedonic responses of the children to the odorants.

However, preferences of older children, 3–7 years of age, in other experiments (Engen, 1974) do, however, show results more similar to those for adults. As has already been stated, the older the child the more similar his or her odor preferences are to those of adults. The most amusing illustration of this developmental difference was seen in the just-mentioned experiment involving the presentation of odorants to children from behind a pictorial display. Even though the mother was seated away from the atomizer, she could detect the odor and would often show signs of displeasure with the unpleasant odorants, in particular dimethyl disulfide, and occasionally would express concern that the child did not seem to respond to it. Some mothers wondered whether their children had a poor sense of smell. The explanation given was that in our experience this was not a surprising result because children can tolerate odors better than adults. This tolerance, or lack of preference, is also obtained in older children, as, for example, when 30% of 4-year old children compared with 2% of adults in a pair comparison (Figure 8.4) chose butyric acid over neroli oil, diethyl phthalate, or rapeseed oil (Engen, 1974). This is not the first time such observations have been made. Peto (1936), a physician, examined 293 children (without any olfactory problems) in a hospital in Budapest and found no evidence of displeasure from children below age 5 to odors found disagreeable by adults and by older children.

Are There Universally Pleasant and Unpleasant Odors?

Many factors affect odor preferences. They vary with the times (men now wear perfumes in the West) and from culture to culture (Ellis, 1928). Moncrieff (1966) writes that although floral perfumes are preferred in the Western world, in the East "perfumes are heavy, intriguing, sleepy and

Are There Universally Pleasant and Unpleasant Odors?

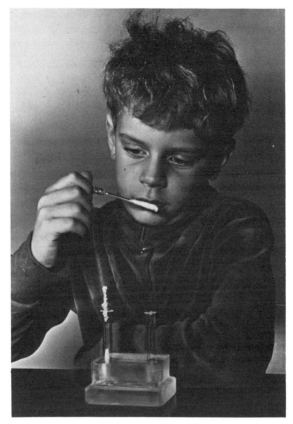

Figure 8.4. Pair-comparison odor-preference testing of a 7-year-old at the Injury Control Research Laboratory, Providence, Rhode Island.

mildly intoxicating [p. 297]." Age may also be an important factor, and so, of course, may gender, although contrary to advertising suggestions that there are uniquely masculine and feminine scents related to exotic body odors, the same perfumes may be sold to men and women under different labels. In general, one's circumstances and experience, habits, and expectations regarding product brands affect such preferences (Moskowitz, 1979a; Cain, 1979b).

Dislikes also vary. In the Western world, some adults complain about the manure of cattle (Lindvall et al., 1973), whereas others do not mind and may enjoy the odor of a barn. However, no one seems to enjoy the odor of the privy or the septic tank, and this may be a universal dislike. Studies have revealed that among offensive odors resulting from cellulose factories and other industries are hydrogen sulfide, methyl mercaptan, dimethyl sulfide, and dimethyl disulfide (a reason we tried them with the children in our studies). Engines—especially those running on diesel fuel—refineries, synthetic resin reactors, printing enamel plants, food-processing plants, soap factories, and garbage dumps also emit odors found unpleasant by most people (Third Karolinska Symposium on Environmental Health, 1970). The most unpleasant odors associated with people and their homes, according to a report from Monsanto (Sellzer, 1975) are proton acceptors or donors, including carboxylic acids in sweat and rancid foods, thiols, phenols, amines, and tobacco smoke. Of course, tobacco smoke seems to be more unpleasant to nonsmokers (including those who have quit) than smokers and children (7–15) also dislike tobacco smoke, though perhaps because of the eye and nasal irritation rather than the odor (Cameron, 1972).

Moncrieff (1966) asked subjects to rate various samples of odors, some expected to be found pleasant, some unpleasant. His adult subjects preferred natural odors from fruits, vegetables, and flowers, such as honeysuckle flowers and fresh parsley, over chemical and synthetic odorants such as amyl acetate, acetone, and camphor. The most disliked odors were again among those listed earlier, plus butyric acid, ethyl mercaptan, and pyridine. Although there are exceptions, Moncrieff believes that the complexity in terms of the number of constituents is an important aspect of odor preference. For example, there are some 25 active ingredients in raspberry. Perfumes are also generally mixtures rather than simple animal or floral scents (Moskowitz, 1978).

Moncrieff also points out that the higher the concentration the less pleasant the odor might be, an observation that has been documented experimentally in psychophysical scaling experiments (Henion, 1971). For neutral and unpleasant odors, their intensity and quality may be negatively correlated attributes of the same subjective dimensions. Pleasant odors such as eugenol and ethyl butyrate, however, may reach an asymptote in pleasantness as concentration in increased and may even become unpleasant at high concentrations (Moskowitz, 1977; Moskowitz, Dravnieks, & Klarman, 1976). Psychological experience probably does not vary in one dimension. Philosophers since Mill have considered that hedonic experiences differ quantitatively in terms of intensity and duration, but also qualitatively in both feelings and desirability (see Edwards, 1975). A stench might come to dominate and motivate all of a person's awareness, but neutral and even pleasant odors may have a different and less dramatic perceptual effect. A reported discovery at Monsanto of a "fresh air" smell that counteracts malodors without affecting pleasant odors could provide a new basis for understanding of pleasant versus unpleasant odors, but the company is filing for a patent and has not yet divulged the secret behind this unusual effect (Sellzer, 1975). The methods and results have also not been without critics (Pantaleoni, 1976).

Do Odors Affect Health?

Many persons seem to believe that odors do affect motivation, and a number of socioepidemiological studies have dealt with the possibility that odors make people sick. Where careful studies have been made in Sweden, complaints about odors range from 27% of people interviewed in rural areas to 78% in urban areas. Similar studies in both Sweden and the United States reveal complaints about the effect of malodors, even at distances more than 20 km from the source of the odors (Cederlöf, Edfors, Friberg, & Lindvall, 1965; Friberg, Jonsson, & Cederlöf, 1960). Shigeta (1971) notes that even though very few industries in the United States were in violation of air pollution regulations, over 30% of the complaints concerned malodors. This might make one suspicious that the respondents are biased and the data not trustworthy. However, there has been a change in the official attitude toward such data. Greater emphasis is now placed on reports of annoyance due to environmental factors (Lindvall, 1969), and such reports are used in intervention by authorities. One is now willing to consider that the human nose may in fact be a better indicator than physical or chemical analysis of pollutant concentrations, especially when the data are obtained by investigators trained in psychophysics (Horstman, Wromble, & Heller, 1971). Mobile laboratories have made it possible to test subjects with air samples of malodors from their own communities (Lindvall, 1970; Springer 1974) to attempt to make the information both psychophysically reliable and socially relevant.

There is, however, no evidence that malodors per se will cause physical disease, but the definition for well-being is also being revised. The World Health Organization now proposes this definition: "A state of complete physical, mental, and social well-being, not merely the absence of disease or infirmity." Thus epidemiological studies may document the effect of purely psychological factors. It is not known how much the purely psychological effects of odors could affect well-being. Some believe they could become intolerable. In discussing the possibility of installing holding tanks for marine toilets to comply with new EPA regulations, one sailor commented:

> The head law is a bad law. Not only does it require a retro-fit, something that would be unheard of if applied to automobiles, but in severe conditions it could actually threaten the lives of boatmen. We need only look back a few months at the events of the Fastnet Race to see how such a thing could happen . . . all of the men who died in the Fastnet Race were people who abandoned boats

that did not sink. One of the prime reasons for leaving their vessels was because it was unsafe to be on deck and unbearable to stay below. In such conditions it does not take much stretch of the imagination to visualize a holding tank breaking loose, or a treatment device rupturing and spilling its contents—creating a stench that would drive most people topside or into a raft [Hammond, 1980].

The Pleasantness of Odor and Bodily State

Cabanac (1971) has shown that the physiological state or "internal milieu" of a hungry person tends to affect his or her hedonic judgments, but not the psychophysical judgments of intensity, which presumably are under the control of external stimuli and thus independent of state. This state primarily affects hedonic responses to food odors. An odor of a food is pleasant when one is hungry and on the unpleasant side after one has overindulged in the food, but the perceived intensity of the odor tends to be unaffected. By contrast, nonfood odors such as laboratory chemicals generally do not show such shifts in preference judgment (Mower, Mair, & Engen, 1977). There may also be exceptions to the hypothesized stability of perceived intensity despite change in bodily states. For example, there is some evidence that one may become more sensitive with deprivation to a substance the body needs, such as vitamins. In one experiment, human subjects were found to be less sensitive or have higher thresholds for thiamine after consuming 10 mg of it compared to control subjects who had received a placebo (Comrey, Klein, & Watson, 1958). Furchtgott and Friedman (1960) reviewed the earlier literature on this topic and did several experiments on their own. They concluded that there is an increase in sensitivity to odor (and taste) if, for example, one skips lunch, but the effect is rather small and may readily be masked by other factors associated with the psychophysical methodology and individual differences. (See also Berg, Pangborn, Roessler, & Webb, 1963; Pangborn, Berg, Roessler, & Webb, 1964).

Need depends on homeostasis and may affect hedonic responses to odors, but, as noted earlier, such likes and dislikes are consequences of smelling and tasting food. Foods that lead to well-being will be liked and those that make one sick or were eaten while one was sick will be disliked. Such experiences can override deprivation and satiety effects (Mower et al., 1977), with profound results. In an electrophysiological recording experiment with rats (Pager, 1977), eucalyptol, which has a camphor-like odor and is not part of laboratory rat food, was added to it. After the rats ate this food, electrophysiological recordings during olfactory stimulation with eucalyptol acquired the ability to activate mitral cells in the olfactory bulb. Mitral cells are believed to serve the role of integrating such qualitative activity in higher centers in the nervous system (LeMagnen, 1971). Pager

(1977) showed that the response from the mitral cells tends to habituate more slowly to repeated presentation of the odorant when the rats are hungry than when satiated.

Individual differences may be larger in judgment of intensity than in judgment of pleasantness, as Degobert (1979b) found in attempting to determine a suitable standard of comparison to be used in evaluating diesel odors. Although pyridine was found to work well for most observers in soliciting comparative judgments of intensity for such odors, no agreed-upon standard for unpleasantness could be found. For example, although most observers found mercaptan repulsive, some liked it because it reminded them of "onion soup," although not the best onion soup, according to Degobert (p. 109)! Individual differences produce an unusually wide distribution of ratings, and some odors are bimodal, according to Land (1979), as in the case of the onion soup, with some persons liking the odor, and others disliking it.

In general there is surprisingly little success attributed to the use of unpleasant odors as deterrents. In aversion therapy in treating alcoholism and overeating, it is necessary to use many unpleasant odors because habituation is apt to set in and thus reduce the initial unpleasant reaction to the odor if it itself has no negative consequences. Mere exposure to an odorant will influence its pleasantness such that a pleasant odor will become less pleasant, and an unpleasant odor more pleasant over time (Cain, 1979b). This principle apparently applies equally to workers in perfumes and candy stores and to those in rendering and glue factories and pigsties. Likewise, there has been success in controlling predatory behavior through association training (Garcia & Brett, 1977), but not in developing dog repellents that are obnoxious to dogs.

Whether odors can make one sick then is not entirely clear. Can they make one feel *good?* Bodily state affects odor perception, but does the opposite happen? It is the purpose of perfumers to affect perception and to enhance well-being (M. A. Mitchell et al., 1964) by creating good odors and masking undesirable ones. Giving an old car a "new car" odor is clearly an attempt to control behavior by taking advantage of associating the old car with a desirable odor. This is even done in the case of relatively odorless products, such as underwear, or plastics made to smell like real leather (Moncrieff, 1966). But here again one takes advantage of prior associations. The question is whether there are inherently meaningful odors as illustrated by the following question.

Is There a Human Pheromone?

In a series of papers, Wiener (1967a,b, 1968a,b) cited over 800 references for the existence of significant sources of human body odor, and argued that the human sense of smell can be much keener than generally assumed. Wiener also argued that there is a strong likelihood of human olfactory communication through odorous "external chemical messengers." Accordingly, humans produce various specific odorants with specific effects on other humans, paralleling pheromonal behavior in animals. The paragon of a pheromone is bombykol, the airborne hormone secreted by the female silk moth, *Bombyx mori,* which has an irresistible control over the male even at distances measured in kilometers (Jacobson, Beroza, & Jones, 1960). Comfort (1971), the noted author of the popular book *Joy of Sex,* also suggested that there are several reasons to expect that human pheromones will be found. These authors counteract the unproven assertion that the human sense of smell is not nearly as acute as that of animals with the argument that it has been repressed, a point also noted earlier in connection with the development of odor preferences in children (Stein et al., 1958).

The first evidence marshalled in support of the existence of a human pheromone is a condition that may be described as hyposmia related to ovulation. Threshold measurements have shown that the olfactory acuity of women is best at that time and poorest during menstruation, thus suggesting that there may be a logical connection between this cycle and the procreation of the species. The phenomenon was first observed in humans by LeMagnen in 1952, and he related it to earlier animal work done in 1950. Since that time many others have made contributions to the topic (Doty, 1977b; Doty, Snyder, Huggins, & Lowry, 1981; Schneider, 1974). There is a relatively large, transient increase in estrogen at the time of ovulation, another fact that helped to stimulate interest in the topic.

The next bit of evidence for the possibility of communication by odors in humans was presented by McClintock (1971), who showed that the menstrual cycle is influenced by the people a woman interacts with. Even as an undergraduate McClintock had begin to explore the old wives' tale that the cycle of women living together would be in synchrony. She demonstrated that this was the case with 135 women living in single or double rooms. The longer women spent time together, the more similar was the time of onset of menstruation for such roommates as well as for close friends, compared with random pairs of women and with controls for other factors, such as diet, which might have influenced the results.

To explore further the hypotheses that this represented a pheromonal effect related to social groupings, women were divided into two groups, those who "spent time with males three or more times per week," versus those who spent less time with the opposite sex. Borderline cases and women taking birth-control pills were excluded from the comparison. Those who estimated seeing males less than three times per week experienced statistically significantly longer cycles than those of the other group, whose mean cycle corresponded with national norms of approximately 28 days. McClintock realizes that exposure to men may not have been the actual cause; it could have been, for example, that for some other reason those women with longer cycles were less likely to spend time with males. However, "one subject reported that she had a cycle length of 6 months until she began to see males more frequently. Her cycle length then shortened to 4.5 weeks. Then when she stopped seeing males as often, her cycle lengthened again [p. 245]."

Russel, Switz, and Thompson (in press) took this research a step further by manipulating the odor women were exposed to and then observing their cycles. They first obtained underarm perspiration from one of the female researchers, collected on a sterile cotton pad. This was diluted in alcohol and daubed below the upper lip and just under the nostrils of eight women, another eight serving as control subjects who were daubed with alcohol only. Then they were observed at regular intervals for 4 months. The control subject's cycles remained essentially unchanged, but the experimental group exposed to the researcher's underarm odorant were on the average only 3.4 days apart in the onset of their cycles compared with an average of 9.3 days at the onset of the experiment. It is as though the cycle of the donor came to "drive" the others.

The most exciting study, according to the popular literature, is one from Great Britain that attempted to assess the effect of odors on social interactions (Cowley, Johnson, & Brooksbank, 1977). It is reported (e.g., *Science News*, April 28, 1979) as though it provided support for a subliminal pheromonal effect in the sexually expected direction. However, it is in fact a very complex social psychological experiment with complicated results, and it can only be said not to reject the hypothesis that human pheromonal effects exist. The experimenters presented two odorants presumed to be pheromones, a mixture of short-chain fatty acids from female primates, "copulin," and androstenol (5α-16-androsten-3α-ol), a natural 19-carbon steroid found in higher concentrations in the urine and sweat of men than in that of women. These odorants were applied to disposable face masks worn by the subjects while interviewing candidates for an important position in the university administration under the pretext that the mask would conceal their facial expressions. Even though the authors state that the students were apparently unaware of the true purpose of the experiment, one wonders how they were affected by the masks and how a control group not wearing the mask would perform. The results suggested that the androstenol affected only women and then only when they were rating male candidates on favorable questions. However, the aliphatic acids had no comparable effect on men, which is surprising considering related work described later in this section. It is possible that any odor might have the same effect that androstenol had for other than pheromonal reasons, because the necessary control groups were not tested. In addition to the difference in gender, there were such large individual differences within each sex that no general conclusion regarding sex or any other hypothesis can be accepted or rejected.

Another experiment with surgical masks (Kirk-Smith & Booth, 1980) found that photographs of women (again more so than men) were judged sexually attractive in the presence of androstenol. In a related experiment (Kirk-Smith, Booth, Carroll, & Davis, 1978) an androstenol aerosol was sprayed on certain seats in a patients' waiting room in a dental clinic. Women tended to choose these seats, and men, surprisingly, tended to avoid them. In both experiments the authors indicate that other than a sexual explanation, including the possibility of prior conditioning to androstenol experience in perfumes, could not be rejected.

The belief in human pheromones was strengthened considerably by evidence presented by Michael and Keverne (1968) that there seemed to be such sex attractants operating on other primates (rhesus monkeys, toque macaques, and bonnet macaques). They proposed the hypothesis on the basis of research on vaginal secretions that short-chain aliphatic or volatile fatty acids used in the study just discussed could be the pheromone operating in the case of the primates, and this substance later became known as "copulin" (Michael, Keverne, & Bonsall, 1971). These investigators then developed a tampon method and collected vaginal samples from young women and found again volatile fatty acids near the middle of the menstrual cycle as they had in case of the primates (Michael, Bonsall, & Warner, 1974). The copulin found in women was reported to possess sex attractant properties in the primates.

Finally, however, still more research on the primates brought out evidence casting serious doubts that copulin was a pheromone. According to Keverne (1976) two basic problems became evident as the research progressed. First, copulin is not a stable stimulus but tends to vary or to be modified by bacteria in the vagina. One of the aspects of bombykol, the model pheromone, is that its chemical nature is precisely defined and it is stable under the circumstances described as an insect attractant. Second, copulin does not seem to control the primates in the same manner as bombykol controls the male moth. The primates seem more variable, showing an evidence of plasticity of behavior, or being less stimulus-bound. The experiments with the primates consisted of operant conditioning of males to press a lever and thereby gain access through a door to a female in the next cage. The female's ovaries had been removed, and the male's performance was observed when she had been given estrogen, and would thus be sexually ready, and when she had not received estrogen, and presumably could not emit copulin, which is under the control of estrogen. The males were also tested both when their nostrils had been plugged with cotton, making them anosmic, and when their sense of smell was unimpaired. The early observations showed that the male would bar-press to gain access to the female only when its sense of smell was unimpaired and when the female had been given estrogen. When the male was anosmic, or in the absence of estrogen for the female, bar-pressing was virtually absent. However, deviation from this kind of stereotypical behavior was observed later when a male might bar-press to gain access to the female even when anosmic and even when the ovariectomized female had not received estrogen. It seems that memory of previous occasions and a certain female with which the male had enjoyed sex earlier was a factor. According to Keverne,

> if we are to consider the complexity and plasticity in the response of the male rhesus monkey, then the further social and cultural evolution of man may make the search for an olfactory aphrodisiac with sexual releasing properties a fruitless task. Indeed, it may be argued that stimulation or provocation by female odours could be disruptive of our social order and perhaps this is why we take such pains to disguise our body odours. This is not of course to say that these odours play no part in human sexual behaviour, but to give them significance at the level of sex attractants underestimates the complexity of human behaviour [p. 266].

Others have also been unable to duplicate the original primate data (Goldfoot, Kravetz, Goy, & Freeman, 1976).

Studies of human vaginal odors show that they change both in pleasantness and intensity during the menstrual cycle (Doty, Ford, Preti, & Huggins, 1975). A study with couples using copulin on the woman's chest as a perfume did not have any apparent effect on the frequency or any other aspect of intercourse (Morris & Udry, 1978). Human body odor from T-shirts worn for 48 hr (without perfume or deodorants) tends to be rated as unpleasant and, except for the rater's own shirt, as being associated with negative attributes such as being unattractive to the opposite sex (Levine & McBurney, 1977; McBurney, Levine, & Cavanaugh, 1977). However, there are exceptions. A person with such body odor might be rated as "active, "strong," "industrious," and "athletic."

There is no doubt joy in sex, but joy probably precedes the sensory pleasures of the odor associated with it. We come back again to the hypothesis that odors become meaningful through experience and association with other events, things, and people.

In animals there is in fact evidence that experience influences the sexual significance of odors (see Nyby, Whitney, Schmitz, & Dizinno, 1978). A perfume was found to stimulate sexual behavior in male mice after they had encountered female mice odorized with it. The results suggest that naturally occurring odorants assumed to be pheromonal may also acquire their sexual significance after such association.

However, biological factors do affect odor sensitivity. There is little doubt that women are more sensitive during ovulation than during menstruation, but more for biological than psychosexual reasons. LeMagnen (1952) believed that the change in the subject's odor sensitivity was specific to biologically significant odorants presumably from the axillary and pubic apocrine glands. He and other investigators directed their attention to Exaltolide, a synthetic must 15-hydroxypentadecanoic acid similar to mammalian substances such as muscone from the musk deer, and used as fixatives in perfumes. Exaltolide is similar to androstenol which was discussed earlier. However, similar sensitivity changes during the menstrual cycle have since been observed for other odorants, such as citral and m-xylene, not believed to be of any biological significance (Köster, 1965; Schneider & Wolf, 1955). We propose that the variation in odor sensitivity during the cycle might have a much simpler explanation than pheromones, namely that it is the result of accessibility of odorants during menstruation due to changes in the mucus level at the olfactory epithelium (Mair et al., 1978). During menstruation there is more mucus present and thus an increased propensity of mucus to absorb molecules and prevent them from reaching the receptors. During ovulation the mucus layer is thinner and thus the accessibility of molecules is greater. This thickness is in turn controlled by hormones, testosterone and estrogen. In other words, what varies during menstruation seems to be the likelihood that molecules will reach the receptors in a way that seems quite consistent with the gas chromatographic model described in Chapter 2. It seems that Exaltolide has relatively large and nonvolatile molecules, which would be more readily affected by this than, for example, an acetate with smaller and more soluble molecules.

In any case, the variation in sensitivity remains an interesting problem. There are also reports that women are more sensitive in general than men, not only during menstruation (Koelega & Köster, 1974). These investigators also report that prepubertal children tend to be less sensitive to "biologically significant odors," as men are to Exaltolide. However they point out that there are controversies regarding the results.

We should also recall the similarity between the olfactory and immunological systems. According to O'Connell (1978), both are chemical senses and may be regulated by the same genetic mechanism associated with the major histocompatibility complex and thus may be involved in pheromone production and perception. Both systems may be involved in discriminating the self from the nonself (Thomas, 1974).

chapter

9

INTERACTIONS OF ODOR AND OTHER PERCEPTIONS

Odor and Taste

One very common observation is that odor affects taste and that what people ordinarily call taste is really flavor, a combination of gustatory, olfactory, and other sensory input. Perception of odor may, therefore, play an important role in appetite, and those who have lost their sense of smell (Chapter 5) may suffer because of this. Obese people are said to be more influenced by such sensory cues than are people of average weight, who respond more to internal factors in regulating food intake (Schachter, 1971).

In the physiological sense, taste refers to stimulation transmitted over the lingual or facial nerve (seventh cranial), the glossopharyngeal nerve (ninth cranial) and vagus nerve (tenth cranial), whereas smell involves the olfactory nerve (first cranial). Chewing and eating food involve all of this input, and, of course temperature and texture. Yet the subjective experience is not a list of sensations, but an integrated flavor or aroma. By simply holding one's nose one can change the characteristic taste of a food dramatically. Even the detection of nonolfactory stimuli such as the sweetness of pear nectar is more acute when subjects' noses are used naturally than when they are blocked off with swimmers' nose clamps (Pangborn, 1966).

The crucial role of perceived odor is easily observed in such a simple demonstration even though pinching one's nose does not prevent all olfactory molecules from reaching the olfactory epithelium, because some of them can get there through the back door of the nasopharynx. To prevent this and make subjects truly functionally anosmic, Mozell and his colleagues (Mozell, Smith, Smith, Sullivan, & Swender, 1969) flowed purified, odorless air through the nose via Teflon nose plugs (illustrated in Figure 9.1), thus preventing odorants from getting through when testing gustation. The flow rate was 2.25 liters/min per naris, which approximates the normal rate of breathing. In order to eliminate differences in texture and temperature, the taste stimuli were liquefied in a blender and presented at the same temperature on the subject's tongue with an eyedropper. One could, therefore, compare the subject's ability to identify the food put on the tongue with and without the contribution of odor perception when purified air was flowed through the nose while tasting. The importance of the odors of the substances was tested separately by flowing deodorized air through each substance, thus reconstituting its odor before it was presented to the subject through the nose plugs. The food substances were coffee, wine, vinegar, cherry, whiskey, lemon, molasses, onion, garlic, apricot, pineapple, cinnamon, clam juice, root beer, chocolate, cranberry juice, grape juice, dill pickle juice, salt water, sugar water, and plain water. Twenty-one sophomore medical students, 18 men and 3 women, served as subjects.

The results showed that when tasting the substances normally without nose plugs the subjects had an average score of about 60% correct identification, with a range of scores from 40 to 95%. When the odorized air was flowed through the nose the results were comparable, indicating that there were no unusual effects associated with the testing procedure. When the odorous molecules were blocked the average score fell to about 10% correct identifications with a range from low to high of 0 to 25. Of course, other subjects, such as housewives and chefs, might have done better, but

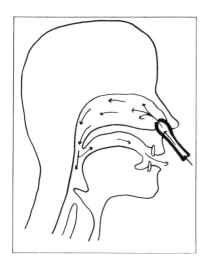

Figure 9.1. Schematic drawing of airflow of the odorant. (From M. M. Mozel, B. P. Smith, P. E. Smith, R. J. Sullivan, Jr., & P. Swender, Nasal chemoreception and flavor identification. *Archives of Otolaryngology*, 1969, *90*, 131–137, Figure 1. Copyright 1969, American Medical Association.)

the concern here is not with such individual differences, but with the ability of *the same* normal subjects to identify food flavors with and without odor perception.

There were also differences in the scores for the various foods, both in how well they were identified in the normal tasting condition, and in the extent to which they were affected by the anosmia. For example, Figure 9.2 shows that 95% of the subjects could identify coffee when they could smell it, but none of them could identify it when they were made anosmic. Similar drastic changes took place in the case of cherry, molasses, garlic, apricot, pineapple, root beer, chocolate, cranberry juice, and dill pickle juice. But even presumably gustatory stimuli such as plain water, salt water, and sugar water were identified less accurately when odor perception was eliminated. It was vinegar, whiskey, and lemon, in particular, that accounted for the correct scores under the functional anosmia, possibly because of greater involvement of the trigeminal nerve.

A Paradox?

Odor and taste are even more closely integrated aspects of flavor than this example indicates, as demonstrated by a related study that observed the effect of varying the amount of odorant and tastant in mixtures scaled for perceived taste intensity (Murphy, Cain, & Bartoshuk, 1977). Sweet-tasting saccharin, which has little if any odor, was added to fruity-smelling ethyl butyrate, which only has a slight taste. Various mixtures of different proportions of these two stimuli were scaled for the perceived taste intensity with the method of magnitude estimation. The perceived intensities of the components of the mixtures were also scaled separately by tasting the saccharin and smelling the butyl acetate. The taste intensities of the mixtures were found to be 93% of the sum predicted from the scale values of the component odorants and tastants. Gas chromatographic analysis provided satisfactory evidence that these results were psychological and not the results of physical changes in the vapors that might have come about from adding saccharin to the ethyl butyrate. Taste intensity was scaled under two conditions, with the nostril open normally, and with the nostril pinched closed by the subject. Blocking the nose made the perceived intensity of the taste of the mixtures weaker but had no appreciable effect on the pure saccharin solution. The greater the concentration of the odorant in the mixture the greater the effect of blocking the nose on the perceived intensity of taste; that is, the taste was weaker. In other words, flavor seems to be described in terms of taste sensation rather than smell. It is not simply that flavor is dominated by odor, as in the example discussed previously, but odor and taste are actually confused.

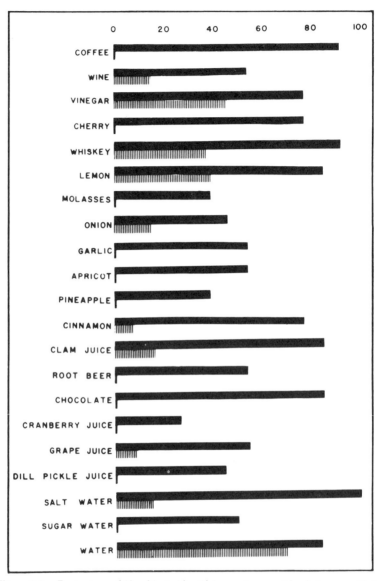

Figure 9.2. Percentage of 13 subjects identifying various sapid substances with (solid bar) and without (shaded bar) odor perception. (From M. M. Mozel, B. P. Smith, P. E. Smith, R. J. Sullivan, Jr., & P. Swender, Nasal chemoreception and flavor identification. *Archives of Otolaryngoloy,* 1969, *90,* 131–137, Figure 3. Copyright 1969, American Medical Association.)

In a follow-up experiment (Murphy & Cain, 1980), taste and odor mixtures were investigated, with special interest given the effect of how well conponents seemed to blend. Mixtures made of citral (which smells like lemon) and salt (sodium chloride, NaCl) seemed to be "dissonant," compared with the more "harmonious" result of mixtures of citral and sucrose, resembling lemonade. Both the dissonant and harmonious mixtures gave results similar to the original results from mixtures of saccharin and ethyl butyrate. There was again evidence that adding and removing odor perception by varying the proportion of the odorant in the mixture affected taste perception. It is apparently not merely a semantic problem peculiar to the present method but a genuine perceptual effect. In other words, when the nose had been closed and then opened, the experience was not that odor was added, but that the substance tasted stronger. Garcia-Medina (1981), made similar observations with mixtures of acetic acid and coffee, which have quite different tastes and odors. Her results agree with those just reviewed, but she suggests, in addition, that evidence that the vector model (Berglund et al., 1976) fits these data "implies that neither taste nor smell contributes to flavor at the expense of the other [p. 13]."

We would like to know what would happen to odor perception with and without taste, although that is probably not a fasible experiment. It would also be instructive to know the subjects' judgment of the flavor in addition to the intensity of the mixtures, because one's mental set influences what is perceived. If a person is asked to identify by label what is tasted, then odor perception seems to play a crucial role. If the set is to judge intensity rather than quality, and one asks the person to judge the intensity of something tasted, then olfactory input seems to play a different role. Although such a psychological decision factor plays a role, explorations of biological factors suggest that there may be more to this than a pure sensory confusion. Palmerino, Rusiniak, and Garcia (1980), in a study of the role of odor versus taste perception in acquired aversions of flavors in animals, suggest that, at least for animals, gustation may provide a better index for such coding of sensory information. Gustation may have a closer association with visceral monitors of the digestive system and may not be as readily interfered with by other sensory stimulation as is olfaction. Olfaction may best be characterized as a busy multipurpose channel used constantly to monitor the environment. They thus propose that olfaction comes to monitor flavor through its association with taste.

Odor and Irritation

Relatively moderate to weak stimulation of the trigeminal (fifth cranial) nerve via free nerve endings in the nasal passages leads to sensations that may also be confused with odor. In one study it was found that 32 of 47 odorants selected as representative of those used in psychophysical experiments in olfaction could be detected by over half of 15 subjects who were anosmic for various reasons (Doty, Brugger, Jurs, Orndorff, Snyder, & Lowry, 1978). The intensity ratings of the anosmics were generally lower than those of control subjects, suggesting that the contribution of the trigeminal nerve to the perceived intensity of "smell" (as some want it called when both the trigeminal and olfactory nerve are involved) can be determined quantitatively.

Cain (1974a) has already obtained such data from two people who, because of acoustic neuroma, had suffered total unilateral destruction of the trigeminal nerve. Those who have suffered this disease will have one nostril with and one without trigeminal sensitivity, and they are, therefore, interesting subjects for an evaluation of the interaction of odor and irritation. And, as expected, magnitude estimation of odor intensities of a range of concentrations of 1-propanol, 1-butanol, and n-butyl acetate revealed that these two subjects perceived the intensity of these odorants to be greater in the "good" nostril than in the one without trigeminal information. Their psychophysical scales are shown in Figure 9.3. On the average the perceived intensity of the affected nostril is about two-thirds that of the "good" one. However, perceived intensity varies with concentration such that at low levels near threshold, no difference in the nostrils can be observed because the trigeminal nerve is less sensitive than the olfactory nerve to such weak concentrations (M. J. Mitchell & R.A.M. Gregson, 1968).

When Cain and Murphy (1980) asked subjects with normal trigeminal nerves to judge the degree of "irritation" as compared with "odor," longer latency was found for the former task. The degree of irritation was found

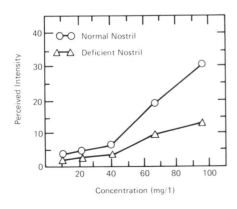

Figure 9.3. Perceived magnitude of intensity with the normal versus deficient nostril in neurectomized subjects. (From Cain, 1974.)

to increase with repeated inhalation. The authors suggest that both the difference in reaction time and the buildup of irritation over time reflect the fact that the nerve endings of the trigeminal system lie deeper in the epithelium than olfactory receptors and thus respond more slowly to stimulation. However, any odorants might stimulate the trigeminal nerve if presented for long enough or at high enough concentration, but the person would not always or even usually know which nerve is involved.

In studies comparing trigeminal versus olfactory stimuli, the distinction has typically been made on the basis of introspective reports from normal subjects using words like *odor* versus *tickling* or *burning*. Such reports are subjective and complicated because they involve semantics, the additional problem of verbal distinctions, which in turn must be interpreted neurologically. Based on electrophysiological data from olfactory stimulation of animals such as the tortoise, Tucker (1971), a physiologist, concluded that "the dream of finding an odorant that is purely olfactory in its stimulating capabilities is still unrealized [p. 170]." It has been suggested that even purified air, depending on its temperature and flow rate, might stimulate the trigeminal nerve (Doty et al., 1978). The problem is an old and complicated one known in psychology in terms of the distinction between so-called adequate and inadequate stimuli. All modalities presumably have their adequate stimuli, such as light and sound to the visual and auditory receptors, but these organs can also be activated by so-called inadequate stimuli. Thus, a hit on the head can make one see stars and hear bells. In general, overstimulation with any stimuli, strong odorants included, can activate the trigeminal nerve and cause pain. This nerve may in fact assume a special role for extremely high concentrations by signaling reflexively an inhibition of inhalation altogether, thus cutting off all stimulation to the nostrils (Alarie, Wakisaka, & Oka, 1973). In the case of subjects having normal trigeminal systems, Cain (1976) asked them, on different occasions, to judge the intensity of the odor, the intensity of the "piquancy" or "burning sensation" (i.e., the trigeminal aspect), and the overall sensation. The results were similar to those he obtained from the patients described earlier in that the intensity of the odor was about two-thirds the overall intensity. The rest could be accounted for as trigeminal. However, the trigeminal effect also increased at a faster rate than the olfactory, as the concentration of the odorant was increased, as shown in Figure 9.4. In other words, odor perception dominates at lower concentrations, but pain and irritation dominate at higher concentrations.

There are stimuli with rather specific effects; for example, carbon dioxide is odorless but stimulates the trigeminal nerve, and perhaps is an example of an ideal adequate stimulus. Perhaps there also are stimuli that stimulate

9. INTERACTIONS OF ODOR AND OTHER PERCEPTIONS

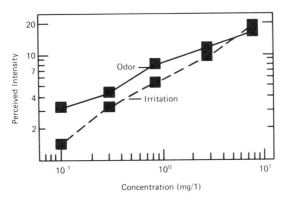

Figure 9.4. Perceived magnitude of intensity of irritation and odor as a function of concentration of n-butyl alcohol. (From Cain, 1976.)

the olfactory nerve only and thereby elicit different odor qualities. However, the search for such stimuli has not been successful (Chapter 2).

Just as taste may inhibit olfaction, as we have seen, so may trigeminal nerve stimulation. Cain and Murphy (1980) showed this by comparing the perceived intensities of various concentrations of an odorant, n-amyl butyrate, and the odorless irritant, carbon dioxide. They tested both physical mixtures of the chemicals prepared before presenting them to the subject as dichorinic mixtures, one chemical to one nostril and one to the other. In both cases, the more carbon dioxide there was in the mixture the more the irritation associated with it would inhibit the fruity odor of n-amyl butyrate, and it did so for both weak and strong overall mixture concentrations. The authors feel that this finding verifies an old hypothesis by Bain (1868, as cited by Cain & Murphy, 1980) that carbon dioxide can arrest odor perception. Support for it is also evident in neurophysiological data showing that carbon monoxide inhibits single units in the cat's olfactory bulb (Hughes & Mazurowski, 1962). In general, "benign" odorants, such as wintergreen, anise, and cherry, were found to facilitate the olfactory bulb activity in unanesthetized monkeys, whereas trigeminal stimuli such as ammonia and acetic acid inhibited such activity. Similar evidence of facilitators and inhibitors had been obtained earlier with psychotic humans with implanted electrodes making possible electrophysiological recording from their olfactory mucosa when stimulated with wintergreen, ammonia, and other stimuli similar to those used in experiments with cats and monkeys (Sem-Jacobsen, Petersen, Dodge, Jacks, Lazarte, & Holman, 1956). Again, one would like to know the effect of varying the concentration of these stimuli, because as we have seen strong stimuli are more likely to be odor inhibitors and weak stimuli odor facilitators.

Odor and Light and Sound

London (1954) cites research in the Soviet Union indicating an interaction and reciprocity between sight and smell. For example, the presence of an odor such as that of pyridine would increase the sensitivity of observers to light. Börnstein (1936) reported analogous results, in a paper quoted frequently even though the description of the experiment is too vague to be replicated and the results can at best be considered inconclusive. In another similar and, typically, isolated experiment (F. Allen & A. Schwartz, 1940), oil of geranium was reported to depress perception of the red part of the visual experience but to enhance the green part; however, the effect of the odor was somehow reversed after the subject had a rest period. In general, individual differences are large in such studies of the effect of stimulation of one modality on the perceptions in another (Stone & Pangborn, 1968).

One extreme case of such modality interaction is synesthesia, in which the stimulation in one modality actually leads to sensation in another modality; for example, a sound may be experienced as a green color. Although synesthesia involving odor is apparently a rare phenomenon, it was reported occasionally by at least one person, a trained psychologist who had been blinded at the age of 11 (Wheeler, 1920). This blind psychologist reported colored tastes and odors, but found them the most difficult to describe of all such experiences. Generally, he described them only as experience of a yellowish and a brownish color for nothing more specific would come to mind. We should not, of course be surprised that odor recall would be difficult, because as noted in Chapter 6 recall of odors at will, in contrast to recognition, is at best a difficult task. Not all people think they can do it and testing for it is a difficult, perhaps impossible, experimental problem.

Although we have noted the possible effect of facilitation, there are no enhancers (Moskowitz, 1978) (such as monosodium glutamate in taste) in olfaction. One might generally think that the presence of more than one stimulus would, in fact, be distracting and thus tend to impair attention and performance in sensory/perceptual tasks. However, a competing hypothesis is that extraneous stimulation actually will arouse the organism and that this in turn will lead to keener sensitivity to sound. The Russian studies mentioned earlier support this hypothesis. Likewise, other studies indicate that people seem to detect sound better with lights on (i.e., arousing stimulation), than in the dark, when there is less distraction (Jamieson, in press; Ozbaydar, 1961).

Noise, too, is arousing and might be assumed to have an arousing and facilitating effect analogous to light in the example mentioned; it should therefore increase odor perception. However, the data available do not support such facilitation of noise on odor perception or vice versa (Berglund, Berglund, & Lindvall, 1975). Noise from road traffic, loud speech in a foreign language (Dutch), music, typewriters, pile drivers, jackhammers, or jet airplanes was present while subjects were engaged in magnitude-estimation scaling of the intensity of odors of various concentrations of hydrogen sulfide. The noises were presented on tape measured to be 57.5, 62.5, and 82.5 dB (A) (by matching them to white noise). On a particular trial, one of these seven sources of noise was presented at one of these loudness levels simultaneously with one of three concentrations of hydrogen sulfide, 1.39, 2.34, and 3.35 parts per billion, with pure air used for comparison. Although there were expected, marked differences in the subjective intensities perceived within each of the two modalities, the results were not affected by the noise. All could be described by the same psychophysical (power) function, which was not affected by the different conditions of exposure to the other modality. However, the authors caution that lack of evidence of such interaction on the purely sensory level measured by a psychophysical scale may be a poor indicator of the effect of noise on comfort, annoyance, or displeasure.

As a corollary to Cabanac's (1971) hypothesis (see Chapter 8), it is conceivable that noise may affect the hedonic dimension of odors without affecting the psychophysical assessment of them. It is apparently only in the case of the closely related sensory modalities of taste and trigeminal stimulation that one sees marked effects of interaction at the purely sensory level. It is, of course, also possible that some other psychophysical approach—for example, measurement of "threshold"—might have revealed an intersensory psychophysical effect, but such a result seems less likely than hedonic effects. In a threshold experiment, M. J. Mitchell and R.A.M. Gregson (1968) asked subjects to judge methanol, ethanol, and 1-propanol by sniffing and tasting, and for irritation by presenting the substances directly to the inner wall of the cheek to maximize stimulation of the nerve endings there. The authors concluded that people can attend to such different sensations from the same stimuli and judge them separately, and that there need be no modality interaction. They do note, nevertheless, that there was suggestion of some such interaction and that the chemical senses are not good examples of sensory independence. A case in point is the example discussed previously in which odor contributes to taste intensity.

Finally, in addition to the sensations discussed here, there are others such as texture, color, and temperature that may likewise interact with one another in the case of foods, but there is much less psychophysical information about them (Moskowitz, 1978). Doty et al. (1978) conclude that "the perceptual scales of intensity, pleasantness, coolness/warmth, and presumed safety are systematically interrelated, with the more intense stimuli being rated as more unsafe, unpleasant and warm [p. 183]."

The Unity of the Senses

According to von Hornbostel (1927), "one must search in order to find the private property of any one sense [p. 84]." His amusing example is that "it matters little through which sense I realize that in the dark I have blundered into a pigsty [p. 83]." Although each sense may be considered a separate channel of information, no sense may make a unique contribution, and again this might be characteristic of the perception of the flavor of foods and the various sense modalities involved. Von Hornbostel goes even further and suggests that the different modalities (e.g., those involved in the pigsty) may be paradoxically equivalent. Gibson (1966), however, argues instead that it is not that different stimuli are perceived to be equivalent, but that they may elicit through association the same information, for example, the perception of a fire may be the same whether the perception is activated by its color, temperature, sound, or odor. In his view the functions of different modalities are under the control of the general activating system of the whole body and represent modes of attention used to explore the information available in light, sound, and chemicals. A pigsty is, of course, a multidimensional stimulus to which one's attention might be drawn by any one of these separate stimuli; the stimulus may in some way lead to the same perception via similar physiological mechanisms, but one would hardly attach the same hedonic value to each of the sensory experiences. Although all modalities may carry information about fires at a cookout and pigsties on farms, one modality somehow seems to characterize the perception, a fire in terms of its brightness or heat, a pigsty in terms of its odor. Some modalities may be more important in monitoring the environment, some in monitoring the viscera (Palmerino et al., 1980).

Those who write about perception have a general tendency to think in terms of cognition. This may not be the best approach for the perception of odor if, as suggested earlier, odor is to emotion what vision is to cognition. When odor is involved it may well cause a feeling before it elicits a concern with the meaning of the odor. According to Achilles' (1929) introspective account, "The first impression is . . . not a pure sensation, as it can be and perhaps often is in vision and hearing, but a complex feeling state. It is the development of this state which takes time, for it entails interactions with other aspects of the situation [p. 276]." This may be the reason that the olfactory sense seems slower to respond than the visual or auditory, although one must conclude so cautiously because there are great differences in reaction measures associated with different testing procedures (Woodworth & Schlosberg, 1954). It is probably not simply that responses to affective stimuli regardless of modality are slower, for Wells (1929) found that people respond just as quickly to pleasant as to umpleasant odors, just as there was no difference in response to pictures of pleasant and unpleasant scenes. The difference seems to be that odors are processed differently. An interesting problem to pursue is whether odors are processed in a parallel as distinguished from serial manner (Sternberg, 1966); whether all the information in short-term odor memory can be examined simultaneously (or in "parallel" fashion) rather than successively or serially, one item after the other. It is technically difficult to measure reaction time to odors, but it is possible that odor perception actually is so rapid that odor-recognition memory is essentially an all-or-none phenomenon. The familiarity of odors does seem to be perceived spontaneously or not at all. What may take time to develop, following Achilles' argument, is the name, the meaning of the odor.

chapter

10

PRACTICAL PROBLEMS AND POTENTIALS OF ODOR PERCEPTION

Some important and interesting perceptual phenomena seem to be peculiar to certain modalities. Illusions, for example, have been thought to be a special problem in vision, but lately clear evidence of analogous phenomena have been demonstrated in hearing (Warren & Warren, 1970). One can replace a phoneme with a clearly audible cough in a sentence, but the perceiver will hear the sentence as though nothing has changed. Besides illusions, constancies and the effect of set are classic topics in the study of perception, but again the examples are almost exclusively visual. Are there similar phenomena in the perception of odors? Consideration of that question involves the veridicality of odor perception, and this leads in turn to another consideration of the value of this modality as an information channel for practical purposes.

Odor Illusions

Slosson (1899) demonstrated the hallucination of odor as a direct result of suggestion. O'Mahony (1978) tried a more elaborate ploy than Slosson's on a British television show dealing with the chemical senses. The viewers were told that there was a connection between the vibration of odorous molecules (see Chapter 2) and the frequency of sound such that the same frequency could be heard and processed by the brain and would lead to the experience of odor. "The brain would recognize these frequencies as smell frequencies and think it must be mistaken about them having come

158   10. PRACTICAL PROBLEMS AND POTENTIALS OF ODOR PERCEPTION

from the ears and that the nose must be actually smelling something [p. 184]." The viewers were told that the odor used in the demonstration would actually be "a pleasant country smell" not likely to be experienced in their homes. They were requested to observe the presentation of the tone and respond whether or not they experienced the odor by calling the station or sending a postmarked letter right away. The number of replies indicated that the suggestion worked although the nature of it was not clear. One hypothesis considered was that the suggestion affected the viewers' response criteria in the sense of signal-detection theory (Chapter 3). The next experiment discussed bears directly on that point.

False Alarms

There is also evidence that the number of false alarms tends to be high in odor perception (Engen, 1972a). A false alarm is a response that a stimulus was present on a trial when it was not. This tendency toward false alarms could reflect poorer stimulus control, variability at the source, or variability in the transmission of odor because of the complex channels of the nostrils. Whereas *hallucination* refers to a sensory experience of something that does not exist, an illusion is more like a misperception of a real stimulus. For example, the moon looks larger at the horizon than at the zenith. Are there illusions involving perceived magnitudes, as in the moon illusion, or restoration of missing elements, as in the speech perception example given? It is not clear if the following example is comparable, but it seems to be primarily a misperception.

In a detection experiment (Engen, 1972), *n*-butyl alcohol was presented on cotton swabs in a standard yes/no detection experiment in which "blanks" of the odorless diluent were also randomly presented. In addition, an odorless coloring agent had been applied to half of the odorant swabs and half of the blank swabs, giving them a slight yellow color. Under one condition of the experiment there was a payoff of one penny for a correct response (hit and correct rejection) and a penalty of one penny for each mistake (false alarm and miss), and under another there was no feedback information about accuracy of judgment at all. Results of the experiment are presented in Table 10.1. It can be seen that on the average the false-alarm rate was higher for tinted stimuli, both with and without payoff. The hit rate was also affected in the same way, but not nearly to the same extent as the false alarm.

After the subjects had completed the experiment, they were interviewed about how they went about judging the odors, in particular whether they had noticed any color characteristic and, if so, what significance they

Table 10.1
Proportions of Hits and False Alarms for Tinted and Clear Cotton Swabs Judged with and without Response Correction on Each Trial[a]

	Stimuli	
	Clear	Tinted
Without correction		
Hit	.91	.91
False alarm	.53	.70
With correction		
Hit	.73	.82
False alarm	.14	.22

[a] From Engen (1972).

attached to it. They all answered that they had noticed the color but had not found it to be correlated with the presence of odor and thus had ignored it! It seemed obvious that a clearly visual and irrelevant stimulus had influenced them unconsciously and decreased their accuracy of performance. This may be classified as an illusion, because false alarms invariably increased with the tint, indicating that the subjects perceived the odor as stronger in its presence. That is, it is not a groundless false perception (hallucination) but a misrepresentation of a real event. Perhaps that distinction is not as important as the fact that it is a real perceptual phenomenon with practical ramifications. It is analogous to perception of odor in the environment when the local factory is emitting variable smoke from its stacks that may or may not be odorous. Reports of detecting odor must be taken seriously, but from a perceptual point of view. A purely physical analysis will be less valid.

Perceptual Constancy

Perceptual constancy exists where, unexpectedly but fortunately, perception remains invariant despite changes in the physical stimulus. This enables the observer to recognize objects and remain oriented. Such examples do not readily come to mind in olfaction, but would include instances where the perception of an odorant remains unchanged, even though the odorant has changed physically or chemically. In speech perception, we hear the same sound (the phoneme "K") in *cow* and *key* even though an acoustic analysis of the sounds made by a speaker shows that it is not physically the same in the two contexts. There are many examples of such constancies in vision, such as form, shape, brightness, and color constancies where stimuli have changed physically without any related change in perception. The color of grass looks the same green under varying physical conditions of illumination affecting the color stimulus.

Although no one has considered the following example in these terms, it could be the explanation for the constancy of the rhesus monkeys' behavior as reported by Keverne (1976). He pointed out that the copulin did not qualify as a pheromone because it underwent chemical change in the vagina, but despite this the animals would continue, in some instances, to behave in the same fashion. Keverne attributes this constancy in performance not to perceptual constancy, but to memory of past pleasures in the same context. One certainly could not reject such a reasonable explanation, but neither can one reject a hypothesis of perceptual constancy according to which the monkeys respond to it as the ame sexually attractive odor. There is no a priori reason to think that olfaction would be different in this respect than other modalities, but problems of this sort demand a better understanding of the psychophysics of odor, both the understanding and description of the stimulus and especially its proximal effect. That is, how constant is the stimulus at the olfactory epithelium? How much can it be changed qualitatively while perception remains the same? How much can one change a rose without affecting its odor?

The only other specific mention of constancies in odor perception is in a paper by Teghtsoonian et al. (1978) on the effect of flow rate on perceived intensity (see Chapter 3). From experiments with olfactometers (e.g., Rehn, 1978) one might expect that greater flow rate would increase the number of molecules reaching the receptors and result in greater odor intensity. However, this does not always happen when the subjects themselves vary the flow rate by the vigor of their sniff; odor intensity seemed to be the same for gentle and strong sniffs in the experiment by Teghtsoonian et al. (1978). The explanation for this, they argue, might be analogous to the constancy of the visual size of an object viewed at different distances giving different sizes of the retinal image of the object. They propose that

> even if flow rate is an important parameter of the proximal stimulus for odor strength (discharge rate in the olfactory nerve), information about sniff vigor may control its effect on the perceptual response. Just as size constancy has the functionally adaptive role of allowing the construction of a stable environment in which objects retain a perceived size independent of viewing distance, so would odor constancy allow the perception of a stable odor source whose perceived strength does not change with sniff intensity. . . . It would be a valuable feature in an odor-detecting system to arrive at the same conclusion about the concentration of the odorant, given either a weakly or strongly inspired sample.

It should also be noted that a critical step in the analysis of size constancy is the determination that it breaks down if the observer is deprived of cues to distance. Thus, for example, given only a Maxwellian view of a target at varying distances, the perception of its size will covary exactly with the resulting changes in retinal image size. If our analogically based concept of odor–strength constancy is valid, we should search for an operation that permits the variation of flow rate while depriving the observer of cues to that variation. Under such conditions our model would predict a clear effect of flow rate on perceived odor strength and would undergo a strong test of its validity [p. 151].

Applications of Odor Research

Training of Odor Perception

It is a widely held view that the human olfactory system might be undergoing atrophy because it is no longer needed and used in modern civilized society. Assuming, correctly or not, that this is so, could the sense of smell be improved with training? Could we all become as good as perfumers are reputed to be? One can safely bet on the ability of humans to learn, and there is evidence that training will be effective in odor perception.

In our early experiments at Brown University we employed the system of measurement proposed by information theory to assess the ability to recognize odors (Engen & Pfaffmann, 1959). We tested the ability of subjects, all untrained college students, to identify odors such as amyl acetate, n-heptanal, n-heptane, and phenylethyl alcohol in terms of their intensities. Each subject was first presented with five stimuli from one odorant in a rank order from weak to strong concentration with a large step between adjacent pairs and asked to learn this rank order by sniffing. When the subject was sure that he or she did know each odor and that further practice would not be necessary, the experimenter would then present each of the five odorants one at a time in a random sequence and asked the subjects to identify its rank order. Each of eight subjects was tested on 250 trials a day for many days and with various odorants. The results were scored in terms of the information measured in bits per stimulus, an index of the extent to which the response matched the odorants on the average (see Attneave, 1959). In the present case, if the subject made no errors the score for five odorants would be 2.32 bits on a perfect matching. The effect of practice was evident from the very beginning. As can be seen in Figure 10.1, which summarizes these data, there was evidence of improvement for eight consecutive sessions or days. The improvement is

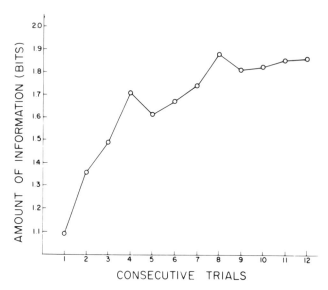

Figure 10.1. The effect of practice by eight subjects on identifying the rank order of the perceived intensity of an odor with sets of five different concentrations of the same odorant. Each trial was with a different odorant. (From T. Engen & C. Pfaffman, Absolute judgments of odor intensity. *Journal of Experimental Psychology,* 1959, *58,* Figure 2 on p. 25. Copyright 1959 by the American Psychological Association. Reprinted by permission.)

marked, nearly doubling in terms of the information measure, and the curve may not have reached an asymptote after 12 days. This improvement was realized simply from correcting the subject's response after each trial, without providing any other information that might make the task easier.

This use of meager feedback and demand for specific labels may have been one reason that we found no such practice effect in our study of the ability to identify different qualities (Engen & Pfaffmann, 1960). However, others (see Chapter 6) have shown that ability to identify odor quality of highly distinct and both extremely pleasant and unpleasant odors improves with specific training (Desor & Beauchamp, 1974). They used "odors emanating from 32 odorous objects or materials," such as coffee, paint, human urine, synthetic sweat, sauerkraut, motor oil, pencil sharpener shavings, cigar butts, and cat feces, in addition to onion, rubbing alcohol, and some more ordinary substances like those we used. Their criterion for a correct response was less strict than ours (presumably including "nearly correct" as well as "precisely correct"), and they used selected subjects, who could identify 60 of the odors after training. This compares with a score of about 16 in the Engen and Pfaffmann (1960) study.

F. N. Jones (1968) earlier also applied what he had hoped would be an effective learning procedure in a similar experiment and with a selection of odorants more similar to that used by Engen and Pfaffmann (1960). During training Jones allowed the subjects to concentrate on doubtful substances, which helped clear up confusion in assigning the proper names to them. Four subjects trained in this way did better than the Engen and Pfaffmann subjects, although not nearly as well as those trained by Desor and Beauchamp. Overall, it has been verified that humans will learn and that noses can be trained.

All these studies show better results than those of Free (1926), who did his study of training noses before any of these others. He found that without any prior training at all an average person, although individual differences were great, could only identify about 1 in 5 or 21.2% in a set of 12 familiar odors such as oil of cloves, hydrogen sulfide, vanilla, acetic acid, and chocolate. These results are close to findings in a similar task of identifying odors by labels, reported by Engen and Ross (1973). The main factor in these tasks is probably that of naming the odor rather than the perception of odor quality per se (see Chapter 6, this volume, and Cain, 1979). It is not simply a matter of transmission of information over the olfactory nerve, as we had assumed earlier, but more of a cognitive task. All of the experiments reported here involved sensory transmission, but not independently of memory storage based on learned association, which is, of course, what the training primarily involves. But, as we have seen, the retrieval of odors involves more than verbal descriptors and may be both affective and cognitive. It is a common experience to encounter an odor with which one is familiar without being able to name it. As noted earlier, we have described this as the tip-of-the-nose phenomenon (Lawless & Engen, 1977a).

Odorous Air Pollution

Psychophysics has been a useful methodology to those doing research in perception since Fechner described it in 1860. A list of practical applications of psychophysical data in vision and audition would be long and impressive. In perception of odor there have not yet been many such accomplishments, but this is changing rapidly (e.g., Cain, 1978b,c). The practical problem of how to measure and define the pleasantness—unpleasantness dimension in connection with combustion toilets represents a recent such accomplishment. Such toilets, which are efficient in vacation homes, have the disadvantage of emitting annoying odors. Lindvall and Svensson (1974) first matched the unpleasantness of the odor of swine manure treated in such toilets to a well-established psychophysical scale of hydrogen sulfide for which perceived intensity is highly correlated with unpleasantness or annoyance. On the basis of quantitative data provided by the psychophysical analysis of the odors, it was then possible to recommend how far apart houses with such toilets ought to be spaced to reduce the odor annoyance. The experiment represents, to date, the only case in which the psychophysics of odor perception has influenced legislation. It is also a significant practical application of psychological scaling in pollution research, which so far has been approached mainly in terms of the threshold concept, which, even assuming it were reliable, could not be expected to predict the experience of suprathreshold stimuli. For example, a threshold concentration of about 10 ppm of hydrogen sulfide has been recommended as a rule-of-thumb guide to what should be permissible in the environment. But to describe all human behavior in terms of a fixed stimulus is not likely to be successful. One needs to think in more realistic terms of the whole range of stimuli and the factors influencing human responses to them (Subcommittee on Hydrogen Sulfide, p. 88).

Artificial Noses

So-called objective olfactometry represents efforts to design an instrument that attempts to model how the nose might transduce odorous molecules into neural responses in the olfactory system and thus might be used for practical purposes. For example, there are instruments to detect gasoline fumes in the bilge of a boat and smoke in a home. Newspaper accounts have described artificial noses designed for the armed forces to detect the enemy through their body odors, principally urine. The main problem with such artificial noses has generally been that they are limited in the number of odorants to which they can respond; they may detect gasoline, or smoke, or urine, but not all three. The fact remains that we do not know what chemical or physical attributes stimulate a response in the olfactory system and thus one has very little information to use in designing artificial noses. Such devices also obviously are not good judges. A person smoking near one such device brought the whole first department to our laboratory. The rumor is that the enemy soon learned to urinate all over the place and thus confused the army's artificial nose. At this stage of research and development, the human judge is very useful, probably essential.

Dravnieks (1968), an expert on the topic of artificial noses and olfaction, states that "To accomplish objective odor measurement at the single substance level, we must; (a) develop a rating system that characterizes sensory or psychophysical parameters of odor, (b) find those instrumentally measurable properties or attributes, (c) find physiochemical techniques that measure the values of the odor relevant attributes, and (d) translate these measurements into values that correspond to the sensory odor parameters [p. 373]."

At this stage of development, artificial noses that can detect different odor qualities, each at a different level of intensity, do not exist, but Dravnieks' work indicates how one might develop such a nose when the psychophysical research has provided the information about the nature of the olfactory detectors one needs to model or duplicate.

Finally, here are some suggestions that may seem farfetched, but in fact are plausible according to clinical observations of animal research.

Diet

It is curious, considering all the schemes for dieting, that none seems to have manipulated food intake through odor control. The data are not definitive, but the indications are that people who become anosmic also have their appetite reduced such that food is no longer as desirable (Schechter & Henkin, 1974). Likewise, it is claimed that food-relevant sensory cues, as distinguished from internal signals of satiety, play a more important role in obese people by stimulating eating (Schachter, 1971). If that is the case, one should be able to reduce the amount of food desired by closing one's nose, or better, by using some nasal spray with a suitable short-term adaptation effect.

Another method would be to attempt to eat one's favorite foods while sick or at the onset of a flu or other discomfort. We have noted that flavors seem especially powerful potential conditioned stimuli for aversion. The typical taste-aversion experiment done with animals is one in which the animal is presented with a novel taste substance (the conditioned stimulus) and at the same time made sick with an injection of lithium chloride (the unconditioned stimulus). One important aspect of such conditioning is that it tends to be selective and to be associated with taste and odor rather than with visual characteristics of food or the place in which the food was eaten, and the unconditioned stimulus nay be present far longer than the conditioned stimulus. This aspect of learning associations can have very profound consequences, as shown by the experiments by Bernstein (1978) that chemotherapy with children is similar to the presentation of an unconditioned stimulus as in the just mentioned paradigm, leading to feelings of nauseousness for many hours during which a regular diet is eaten. The consequence may be dislike of all foods and possibly anorexia nervosa. Most people do in fact have an aversion to certain foods because of associations with feeling sick, either because of overindulgence, food poisoning, or eating the food when an illness was coming on. One acquaintance still thinks bourbon smells like vomit, years after an experience of overindulging in bourbon.

It certainly is the case that one needs to observe what children eat and should not leave such associations and thus food preferences to chance. The time to give children junk food and candy is when they are sick, certainly not when they are healthy and hungry. To leave the development of food habits to children sets the stage for the likelihood that junk food will become associated with feeling good. Good tastes are bred, not born. Indeed, drug tolerance and addictions may be explained on a similar basis (Siegel, 1979).

Pregnancy Prevention

One of the fascinating effects of odor in animal behavior is known as the Bruce effect (Bruce, 1960), which refers to the observation that a female mouse that has been made pregnant by one male tends to abort when the father is replaced by another male in their cage. Also, the pregnancy block becomes more likely if the new male is a member of another strain (Dominic, 1969). Earlier Whitten (1956) had shown that the introduction of a male into a group of females causes the onset of the estrous cycle. The male's presence also influences the length of the cycle. These data and the human data by McClintock (1971) discussed earlier show striking similarities, and the stimulus appears to be olfactory external chemical messengers (Wiener, 1966). Could one bottle the odor of a man and use it the next morning to block pregnancy caused by another man?

A group of Swedish investigators (Bergquist, Nillius, & Wide, 1979) have shown that daily sniffs of a chemical analogue to a luteinizing hormone–releasing hormone (LRH) inhibited ovulation in all but 2 of their 27 regularly menstruating women who served as subjects in their experiment. No mention is made of odor perception in this case but the substance, the LRH analogue, said to have specific pituitary gonadotrophic action and to release of steroid sex hormones, was administered nasally. *Science News* (August 25, 1979, p. 133) describes this research under the heading of "A Sniff a Day Keeps Pregnancy Away." Although LRH analogues have actually been developed with the intention of using them to correct infertility, in this case the analogue paradoxically had the opposite effect. The lesson is to be careful in attempting to apply findings in this area, but there clearly is a connection between the olfactory system and other systems of the body. This may not, of course, have anything to do with olfaction. The nostrils may simply provide access to the blood supply via its capillaries. However, it is reminiscent of the possible relation between immunology and olfaction referred to earlier.

Perfumes

Successful marketing of perfumes may primarily be a matter of teaching people desirable associations, but that topic is beyond the scope of this book. However, we would like to end with a brief mention of the possible involvement of pheromones. We have noted that if there are such chemical messengers they have not been identified as yet, but the research continues. The perfume industry apparently expects to take advantage of sweat or other substances produced by the skin where the perfume is worn by a person. According to advertisements this will produce a perfume that is unique because each person's own ingredients are unique (or at least vary from individual to individual). These personal excretions are apparently similar to the animal products such as musk, which, as noted earlier, is believed to be sexually attractive and thus like a pheromone. Thus, as Ellis (1928) wrote, "we do not really leave the sexual sphere by introducing artificial perfumes. The perfumes which we extract from natural products, or, as is now frequently the case, produced by chemical synthesis, are themselves either actually animal sexual odors or allied in character or composition to the person odors they are used to heighten or disguise [p. 91]."

Musk has been used for a very long time; it has an ancient origin and a Persian name. It is, Ellis suggests, the odor among natural perfumes that most nearly approaches the odor of sexual secretions. Although this may be farfetched, it is important to note that although perfume preferences vary for different cultures and for different times (Moncrieff, 1966), musk seems to be a universally selected perfume ingredient. Could marketing alone explain that? Does musk really have a direct nonlearned physiological effect, or is it that one exposure could be enough to assure musk a pleasant rating never to be forgotten? (See Chapters 6 and 11.)

chapter 11

EPILOGUE

Incidental Associations

The central idea of this book is that the sense of smell is shaped by experience; that is, odors become meaningful through association with other events. This thesis takes issue with the traditional view that responses to odor, especially likes and dislikes, are governed by innate mechanisms. For example, it seems to be a widely held opinion that what is good for the body nutritionally will smell and taste good and what is bad will not, so that if one should find oneself lost in the wilderness with nothing to eat one could use such sensory criteria in searching for nourishment. Sensations presumably provide meaningful information without any prior experience or information. The present view is that odor preferences are acquired by learning to adapt to the environment. Thus, what is innately given is a propensity to learn to associate odor sensations with related negative and positive consequences. Such a learning approach would also argue against the existence of human pheromones or so-called biologically significant odors. Although a popular idea, there is no firm evidence for human pheromones, nor for the existence of a vomeronasal organ which may be the basis of such a mechanism in animals.

Olfactory responses to odorants may *seem* to be innate because they are so readily learned and so long-lasting. One may acquire an odor association by contiguity in just one trial, yet the odor per se may only be an incidental or nonfunctional attribute of a situation, as when overindulgence leads to a lifelong aversion for a perfectly wholesome food.

There is no evidence that the odor itself affects one's health or well-being. The same hypothesis can be applied to biologically significant body odors. One pleasant experience contiguous with the odor of musk, for example, might impart upon it a pleasure to be enjoyed later even though it was not the odor of musk that caused the pleasure. In any case, not everybody enjoys the odor of musk. Sone are indifferent, some dislike it, possibly because of their different previous experiences associated with it. The hypothesis that there are human pheromones may be more intriguing and possibly consistent with an animal model of human behavior, but human individual differences in odor preferences cannot be denied, nor can the viability of an alternative and competing hypothesis of learning by association. To explain why some would dislike the odor of a "pheromone," an odor with inherent stimulus properties, would require a Freudian hypothesis of reaction formation, which has indeed been proposed but not confirmed empirically.

Odor Memory

The data on odor memory also seem consistent with a learning interpretation. There is not enough information to compare the excellence of memory in different modalities in any detail, because one lacks controls for how well associations are made in different modalities based on the same opportunity to learn. However, one noticeable aspect of odor memory is its resistance to the influence of retroactive interference—that is, the effect of subsequent experiences with an odor causing forgetting or interference with its ability to bring forth memories of an earlier association with it. Even relatively meaningless and difficult to name odors encountered in laboratory experiments are recognized after long intervals, during which a person would probably have had many similar odor experiences that one might expect to interfere with memory.

Although interference or inhibition is normally discussed in terms of how it causes forgetting and/or makes it more difficult to learn new material, it can just as well be considered in terms of how it helps to protect old information or associations. It is this latter aspect that seems to characterize odors associated with personally significant memories. Though odor associations seem impervious to retroactive interference, they are affected by proactive interference—that is, the inhibiting effect of old associations on attempts to form new associations with the same stimuli. Therefore, once established, it is difficult to modify odor memories, including aversions. This is of course important for survival, as with the ability to recognize harmful situations encountered earlier, and the point is that such encounters seem readily associated with perception of odors. In fact, at times the persistence of such odor associations actually seems irrational, as when one cannot overcome the aversion to the odor of a perfectly good food one overindulged in long ago, or just happened to have consumed while becoming ill for an unrelated reason. As has been stressed, the odor need not be a functional factor in the situation but may only be a concomitant experience associated with it by contiguity.

Odor memory does seem specialized for recognition and arousal, but the ability to recall odors is poor or nonexistent and certainly not at all comparable with visual and auditory recall. Odors have little if any imagery value that one can re-create, rehearse, or otherwise manipulate mentally.

Habituation and Arousal

Perception of odors is a matter of responding to present, external stimuli. According to prevalent textbook discussions, one can only smell even a moderately strong odor for a matter of minutes before it disappears because of adaptation or receptor fatigue. However, this popular conclusion is based on old experiments and debatable data, on the one hand, and a confusion of adaptation with habituation on the other. Habituation refers to the cessation response, as distinguished from sensitivity, to a stimulus that no longer is considered significant. Habituation commonly happens in a case of odor perception and is a form of learning or adjustment. Adaptation does, of course, affect sensitivity to odors such that a weak odor will be more difficult to perceive after prior exposure to a strong one. It is even possible to put the olfactory system out of function altogether with extreme overstimulation. But the point here is that this is not what typically happens in eating food and exploring the environment. The effect of adaptation may be more pronounced in olfaction than in vision, although no data are available permitting such a conclusion, but light adaptation does not cause blindness, nor does normal exposure to odors cause anosmia. A perfume may be effective all day long.

The effect of cross-adaptation is also likely to be exaggerated. There is no evidence that smoking will make one less sensitive to all odorants any more than a cocktail, or two, before dinner will significantly decrease the sensitivity to all tastes. Rather it seems to be the case that the smell functions as a constant and efficient monitor of the environment.

However, part of this efficiency is that an odorant will be ignored once it has been determined that it denotes nothing of significance. Such habituation may appear as a lack of sensitivity unless tested with psychophysical methods that take the effect of attention and response bias into account. One may not hear an alarm clock unless one needs to pay attention to it, and one may begin to sniff the food if one suspects that it is spoiled. Any novel stimulus will initially get attention but will fade away from consciousness if found to be inconsequential. Yet the sense of smell continues to function without constantly intruding on consciousness, it is always ready to arouse attention.

The Coding of Odor Perception

It is primarily the quality of odor and its hedonic meaning that dominates odor perception. There is a wide hedonic range for odors, but it is generally the unpleasant odors that are effective in arousing conscious awareness. Earlier studies of how odors are coded were preoccupied with attempts to classify perceived odor qualities in terms of a few labels or names, analogous to the use of a few color names in color perception. Odor classification has not increased our understanding of information processing in olfaction. The way people describe odors is quite idiosyncratic compared with their description of colors, because description of odors is influenced more by individual experiences than by inherent neurophysiological processes. In general, vision (or audition) may not be a good model for olfaction. Odor perceptions are evidently assimilated into broader contexts psychologically and physiologically and may be more similar to the perception of pain, especially regarding the importance of the pleasures of sensations, or emotional and motivational factors. As in the case of vision, one can of course identify olfactory sensory characteristics and corresponding neurophysiological processes, but for the reasons outlined earlier interaction with other processes may be an inherent part of odor perception.

REFERENCES

Abrahamsen, A. A. *Child language. An interdisciplinary guide to theory and research.* Baltimore: University Park Press, 1972. P. 162.
Achiles, J. D. Geruchstudien *Archivs für Psychologie,* 1929, *71,* 273–337.
Adams, R. G., & Crabtree, N. Anosmia in alkaline battery workers. *British Journal of Industrial Medicine,* 1961, *18,* 216–221.
Adrian, E. D. The electrical activity of the mammalian olfactory bulb. *Electroencephalography and Clinical Neurophysiology,* 1950, *2,* 277–388. (a)
Adrian, E. D. Sensory discrimination. With some evidence from the olfactory organ. *British Medical Bulletin,* 1950, *6,* 330–333. (b)
Adrian, E. D. The mechanism of olfactory stimulation in the mammal. *Advances in Science,* 1953, *9,* 417–420.
Adrian, E. D. Basis of sensation: Some recent studies of olfaction. *British Medical Journal,* 1954, *1,* 287–290.
Aharonson, E. F., Menkes, H., Gurtner, G., Swift, D. L., & Proctor, D. F. Effect of respiratory airflow rate on removal of soluble vapors by the nose. *Journal of Applied Physiology,* 1974, *37,* 654–657.
Alarie, Y., Wakisaka, I., & Oka, S. Sensory irritation by sulfur dioxide and chlorobenzilidene malontrile. *Environmental Physiology and Biochemistry,* 1973, *3,* 53–64.
Allen, F., & Schwartz, A. The effect of stimulation of the senses of vision, hearing, taste, and smell upon the sensibility of organs of vision. *Journal of General Physiology,* 1940, *24,* 105–121.
Allen, W. J. Effect of ablating the frontal lobes and occipito-parieto-temporal (excepting pyriform) areas on positive and negative olfactory conditioning. *American Journal of Physiology,* 1940, *128,* 754–771.
Allen, W. J. Effect of ablating the pyriform–amygdaloid areas on positive and negative olfactory conditioned reflexes and conditioned olfactory differentiation. *American Journal of Physiology,* 1941, *132,* 81–92.
Allison, A. C., & Warwick, R. T. Quantitative observations on the olfactory system of the rabbit. *Brain,* 1949, *72,* 186–197.

REFERENCES

Amoore, J. E. Odor blindness as a problem in odorization. *American Gas Association. Operating Section,* 1968 (annual), 242–247.

Amoore, J. E. *Molecular basis of odor.* Springfield, Ill.: Charles C. Thomas, 1970.

Amoore, J. E. Olfactory genetics and anosmia. In L. M. Beidler (Ed.), *Handbook of sensory physiology.* Vol. IV. *Chemical senses 1: Olfaction.* New York: Springer-Verlag, 1971. Pp. 245–256.

Amoore, J. E. Specific anosmia and the concept of primary odors. *Chemical Senses and Flavor,* 1977, *2,* 267–281.

Amoore, J. E., Johnston, J. W., Jr., & Rubin, M. The stereochemical theory of odor. *Scientific American,* 1964, *210*(2):42–49.

Aronsohn, E. Experimentelle Untersuchungen zur Physiologie des Geruchs. *Archivs für Anatomie und Physiologie, Physiologische Abteilung,* 1886, 321–357.

Attneave, F. *Applications of information theory in psychology: A summary of basic concepts, methods, and results.* New York: Holt, 1959.

Baker, R. A. Response parameters including synergism–antagonism in aqueous odor measurement. *Annals of the New York Academy of Sciences,* 1964, *116,* 495–503.

Bartley, S. H. *Principles of perception* (2nd ed.). New York: Harper & Row, 1969.

Beauchamp, G. K., Doty, R. L., Moulton, D. G., & Mugford, R. A. The pheromone concept in mammalian communication: A critique. In R. L. Doty (Ed.), *Mammalian olfactory, reproductive processes and behavior.* New York: Academic Press, 1976. Pp. 143–160.

Beauchamp. G. K., & Maller, O. The development of flavor preferences in humans: A review. In M. R. Kare & O. Maller (Eds.), *The chemical senses and nutrition.* New York: Academic Press, 1977. Pp. 291–311.

Bedicheck, R. *The sense of smell.* New York: Doubleday, 1960.

Bedney, B. Mosquitos. In F. D. Reeve (Ed.), *Great Soviet short stories.* New York: Dell, 1962. Pp. 37–71.

Beets, M. G. J. Odor and molecular structure. *Olfactologia,* 1968, *1,* 77–92.

Beets, M. G. J. The molecular parameters of olfactory response. *Pharmacological Reviews,* 1970, *22,* 1–34.

Beidler, L. M. Facts and theory on the mechanism of taste and odor perception. In J. H. Mitchell, Jr. (Ed.), *Chemistry of natural food flavors.* Chicago: Quartermaster Food and Container Institute for the Armed Forces, 1957. Pp. 7–47.

Beidler, L. M. Comparison of gustatory receptors, olfactory receptors, and free nerve endings. *Cold Spring Harbor Symposium on Quantitative Biology,* 1965, *30,* 191–200.

Bekesy, G. von. Olfactory analogue to directional hearing. *Journal of Applied Physiology,* 1964, *19,* 369–373.

Benignus, V. A., & Prah, J. D. Flow thresholds on nonodorous air through the human naris as a function of temperature and humidity. *Perception and Psychophysics,* 1980, *27,* 569–573.

Bennett, M. H. The role of the anterior limb of the anterior commissure in olfaction. *Physiology and Behavior,* 1968, *3,* 507–515.

Berg, H. W., Pangborn, R. M., Roessler, E. B., & Webb, A. D. Influence of hunger on olfactory acuity. *Nature,* 1963, *197,* 108.

Berglund, B., Berglund, U., Ekman, G., & Engen, T. Individual psychophysical functions for twenty-eight odorants. *Perception and Psychophysics,* 1971, *9,* 379–384.

Berglund, B., Berglund, U., Engen, T., & Ekman, G. Multidimensional analysis of twenty-one odors. *Scandanavian Journal of Psychology,* 1973, *14,* 131–137.

Berglund, B., Berglund, U., Engen, T., & Lindvall, T. The effect of adaptation and odor detection. *Perception and Psychophysics,* 1971, *9,* 435–438.

Berglund, B., Berglund, U., Jonsson, E., & Lindvall, T. On the scaling of annoyance to environmental factors. *Report from the Psychological Laboratories, University of Stockholm,* 1975, No. 451. 10 pp.

Berglund, B., Berglund, U., & Lindvall, T. Perceptual interaction of odors from a pulp mill. In *Proceedings of the Third International Clean Air Congress.* Düsseldorf: VDI-Verlag, 1973. Pp. A40–A43.

Berglund, B., Berglund, V., & Lindvall, T. Measurement of rapid changes of odor concentration by a signal-detection approach. *Journal of Air Pollution Control Association,* 1974, 24, 162–164. (a)

Berglund, B., Berglund, V., & Lindvall, T. A psychological detection method in environmental research. *Environmental Research,* 1972, 7, 342–353. (b)

Berglund, B., Berglund, V., & Lindvall, T. On perceptual interaction of noise and odor. *Reports from the Department of Psychology, University of Stockholm,* May 1975, No. 445.

Berglund, B., Berglund, V., & Lindvall, T. Psychological processes of odor mixtures. *Psychological Review,* 1976, 83, 432–441.

Berglund, B., Berglund, U., & Lindvall, T. Psychophysical scaling of odorous air pollutants. Proceedings of the Fourth International Clean Air Congress, Tokyo, May 16–20, 1977. *Japanese Union of Air Pollution Prevention Association,* 1977, 377–380.

Berglund, B., Berglund, V., & Lindvall, T. Olfactory self- and cross- adaptation: Effects of time of adaptation and perceived odor intensity. *Sensory Processes,* 1978, 2, 191–197. (a)

Berglund, B., Berglund, V., & Lindvall, T. Separate and joint scaling of n-butanol and hydrogen sulfide. *Perception and Psychophysics,* 1978, 23, 313–320. (b)

Berglund, B., Berglund, V., Lindvall, T., & Svensson, L. T. A quantitative principle of perceived intensity summation in odor mixtures. *Journal of Experimental Psychology,* 1973, 100, 29–38.

Berglund, B., & Lindvall, T. *Olfactory evaluation of indoor air quality.* Copenhagen: Danish Building Research Institute, 1979.

Bergman, P., & Escalona, S. K. Unusual sensitivities in very young children. *Psychoanalytic Study of the Child,* 1949, 3–4, 333–352.

Bergquist, C., Nillius, S. J., & Wide, L. Intranasal gonadotropin-releasing hormone agonist as a contraceptive agent. *Lancet,* 1979, 2, 215–216.

Bernstein, I. L. Learned taste aversions in children receiving chemotherapy. *Science,* 1978, 200, 1302–1303.

Bienfang, R. Dimensional characterization of odors. *Chronica Botanica,* 1941, 6, 249–250.

Bienfang, R. *The subtle sense.* Norman: University of Oklahoma Press, 1946.

Blakeslee, H. F. Unlike reactions of different individuals to fragrance in verbena flowers. *Science,* 1918, 48, 298–299.

Bobrow, N. A., Money, J., & Lewis, V. G. Delayed puberty, eroticism, and sense of smell: A psychological study of hypogonadotropism, osmatic and anosmatic (Kallmann's syndrome). *Archives of Sexual Behavior,* 1971, 1, 329–344.

Boelens, H. Relationship between the chemical structure of compounds and their olfactive properties. *Cosmetics and Perfumery,* 1974, 89, 1–7.

Bolinger, D. *Aspects of language,* New York: Harcourt, Brace, and World, 1968.

Boring, E. G. *Sensation and perception in the history of experimental psychology.* New York: Appleton-Century, 1942.

Börnstein, W. On the functional relations of the sense organs to one another and to the organism as a whole. *Journal of General Psychology,* 1936, 15, 117–131.

Brill, A. A. The sense of smell in the neuroses and psychoses. *Psychoanalytic Quarterly,* 1932, 1, 7–42.

Bronshtein, A. A., & Minor, A. V. Significance of cilia and their motility in the functioning of olfactory receptors. *Doklady Biological Science,* 1973, *213,* 557–559.
Brown, K. S., Maclean, C. M., & Robinette, R. R. Sensitivity to chemical odors. *Human Biology,* 1968, *40,* 456–472.
Brown, K. S., & Robinette, R. R. Ability to smell solutions of potassium cyanide: Lack of a simple pattern of inheritance. *Nature,* 1967, *215,* 406–408.
Brown, R. *Words and things: An introduction to language.* New York: The Free Press, 1958.
Brown, R. W., & McNeill, D. The "tip-of-the-tongue" phenomenon. *Journal of Verbal Learning and Verbal Behavior,* 1966, *5,* 325–337.
Bruce, H. M. A block to pregnancy in the mouse caused by proximity of strange males. *Journal of Reproduction and Fertility,* 1960, *1,* 96–103.
Cabanac, M. Physiological role of pleasure. *Science,* 1971, *173,* 1103–1107.
Cain, W. S. *Olfactory adaptation and direct scaling of odor intensity.* Unpublished doctoral dissertation, Brown University, 1968.
Cain, W. S. Odor intensity after self-adaptation and cross-adaptation. *Perception and Psychophysics,* 1970, *7,* 272–275.
Cain, W. S. Contribution of the trigeminal nerve to perceived odor magnitude. *Annals of the New York Academy of Sciences,* 1974, *237,* 28–34. (a)
Cain, W. S. Perception of odor intensity and the time-course of olfactory adaptation. *ASHRAE Transactions,* 1974, *80,* 53–75. (b)
Cain, W. S. Odor intensity: Mixtures and masking. *Chemical Senses and Flavor,* 1975, *1,* 339–352.
Cain, W. S. Olfaction and the common chemical sense: Some psychophysical contrasts. *Sensory Processes,* 1976, *1,* 57–67.
Cain, W. S. Bilateral interaction in olfaction. *Nature,* 1977, *268,* 50–52. (a)
Cain, W. S. Differential sensitivity for smell: "Noise" at the nose. *Science,* 1977, *195,* 796–798. (b)
Cain, W. S. Odor magnitude: Coarse versus fine grained. *Perception and Psychophysics,* 1977, *22,* 545–549. (c)
Cain, W. S. History of research on smell. In E. C. Carterette & M. P. Friedman (Eds.), *Handbook of perception,* Vol IVA. *Tasting and smelling.* New York: Academic Press, 1978. Pp. 197–229. (a)
Cain, W. S. Minimum rates of air intake into buildings: Role of odors. In J. A. J. Stolwijk (Ed.), *Energy conservation strategies in buildings: Comfort, acceptability and health.* New Haven: John B. Pierce Foundation, 1978. Pp. 53–70. (b)
Cain, W. S. The odoriferous environment and the application of olfactory research. In E. C. Carterette and M. P. Friedman (Eds.), *Handbook of perception.* Vol. VIA. *Tasting and smelling.* New York: Academic Press, 1978. Pp. 277–304. (c)
Cain, W. S. Interactions among odors, environmental factors and ventilation. In P. O. Fanger & O. Valbjörn (Eds.), *Indoor climate.* Copenhagen: Danish Building Research Institute, 1979. Pp. 257–269. (a)
Cain, W. S. Lability of odour pleasantness. In J. H. A. Kroeze (Ed.), *Preference behaviour and chemoreception.* London: Information Retrieval Limited, 1979. (b)
Cain, W. S. To know with the nose: Keys to odor identification. *Science,* 1979, *203,* 467–470. (c)
Cain, W. S., & Drexler, M. Scope and evaluation of odor counteraction. *Annals of the New York Academy of Sciences,* 1974, *237,* 427–439.
Cain, W. S., & Engen, T. Olfactory adaptation and the scaling of odor intensity. In C. Pfaffmann (Ed.), *Olfaction and taste, III.* New York: Rockefeller University Press, 1969.

References

Cain, W. S., & Johnson, F., Jr. Lability of odor pleasantness: Influence of mere exposure. *Perception,* 1978, *7,* 459–665.
Cain, W. S., & Murphy, C. L. Interaction between chemoreceptive modalities of odour irritation. *Nature,* 1980, *284,* 255–257.
Cameron, P. Children's reactions to second-hand tobacco smoke. *Journal of Applied Psychology,* 1972, *56,* 171–173.
Cederlöf, R., Edfors, M.-L., Friberg, L., & Lindvall, T. Determination of odor thresholds from fuel gases from a Swedish sulfate cellulose plant. *Tappi,* 1965, *48,* 39–48.
Cederlöf, R., Friberg, L., Jonsson, E., Jonsson, E., Kaij, L., & Lindvall, T. Studies of annoyance connected with offensive smell from a sulphate cellulose factory. *Nordisk Hygienisk Tidskrift,* 1964, *45,* 39–48.
Cederlöff, R., Jonsson, E., and Sörenson, S. On the influence of attitudes to the source on annoyance reactions to noise. A field experiment. *Nordisk Hygienisk Tidskrift,* 1967, *48,* 46–69.
Chalke, H. D., Dewhurst, J. R., & Ward, C. Loss of sense of smell in old people. *Public Health, London,* 1958, *72,* 6, 223–230.
Cheal, M. Social olfaction: A review of the ontogeny of olfactory influences on vertebrate behavior. *Behavioral Biology,* 1975, *15,* 1–25.
Cheesman, G. H., & Townsend, M. J. Further experiments on the olfactory thresholds of pure substances, using the "sniff-bottle method." *Quantitative Journal of Experimental Psychology,* 1956, *8,* 8–14.
Chomsky, N. *Syntactic structure.* The Hague: Mouton, 1957.
Clark, E. C., & Dodge, H. W. Effect of anosmia on the appreciation of flavor. *Neurology,* 1955, *5,* 671–674. (a)
Clark, E. C., & Dodge, H. W. Extraolfactory components of odor. *American Medical Association Journal,* 1955, *159,* 1721–1724. (b)
Clark, E. C. Anosmic gustation. *New England Journal of Medicine,* 1968, *278,* 337.
Cohen, H. Clinical and diagnostic significance of the sense of smell. *American Physician,* 1923, *28,* 315–318.
Comfort, A. Likelihood of human pheromones. *Nature,* 1971, *230,* 432–433.
Comrey, A. L., Klein, E., & Watson, G. The effect of thiamine on thiamine sniff thresholds. *Journal of General Psychology,* 1958, *59,* 105–109.
Corbit, T. E., & Engen, T. Facilitation of olfactory detectors. *Perception and Psychophysics,* 1971, *10,* 443–436.
Cornwell, C. A. Golden hamster pup adapts to complex rearing odors. *Behavioral Biology,* 1975, *14,* 175–188.
Cowley, J. J., Johnson, A. L., & Brooksbank, B. W. L. The effect of two odorous compounds in an assessment-of-people test. *Psychoneuroendocrinolgoy,* 1977, *2,* 159–172.
Crocker, E. C., & Henderson, L. F. Analysis and classification of odors. *American Perfumer and Essential Oil Review,* 1927, *22,* 325–327.
Crossland, H. R., Goodman, M., & Hockett, A. Anosmia and its effects upon taste perception. *Journal of Experimental Psychology,* 1926, *9,* 398–408.
Daly, C. D., & Senior White, R. Psychic reactions to olfactory stimuli. *British Journal of Medical Psychology,* 1930, *10,* 70–87.
Davies, J. T. Olfactory theories. In L. M. Beidler (Ed.), *Handbook of sensory physiology.* Vol. IV. *Chemical senses 1: Olfaction.* New York: Springer-Verlag, 1971. Pp. 322–350.
Davis, R. G. Acquisition of verbal associations to olfactory stimuli of varying familiarity and to abstract visual stimuli. *Journal of Experimental Psychology: Human Learning and Memory,* 1975, *104,* 134–142.

Davis, R. G. Acquisition of verbal associations to olfactory and abstract visual stimuli of varying similarity. *Journal of Experimental Psychology: Human Learning and Memory*, 1977, *3*, 37–51.

Davis, A. G. The microencapsulation of odorants as a method of stimulus control in studies of odor quality perception. *Chemical Senses and Flavor*, 1979, *4*, 191–206. (a)

Davis, R. G. Olfactory perceptual space models compared by quantitative methods. *Chemical Senses and Flavor*, 1979, *4*, 21–33. (b)

Degobert, P. Sensory behaviour of a nose panel towards bad odours: Application to diesel exhaust odours. In J. LeMagnen and P. MacLeod (Eds.), *Olfaction and taste*, VI. London: Information Retrieval Limited, 1977. Pp. 459–470.

Degobert, P. Diesel exhaust odours, their importance—Evaluation methods. *Institut Français du Petrole*, Ref. IFP, No. 26864 A, April 1979. (a)

Degobert, P. Hedonic and intensity ranking of different malodors by category estimation and pain comparison. In J. H. A. Kroeze (Ed.), *Preference behaviour and chemoreception*. London: Information Retrieval Limited, 1979. Pp. 107–121. (b)

Desor, J. A., & Beauchamp, G. K. The human capacity to transmit olfactory information. *Perception and Psychophysics*, 1974, *16*, 551–556.

De Vries, H., & Stuiver, M. The absolute sensitivity of the human sense of smell. In W. A. Rosenblith (Ed.), *Sensory communication*. New York: Wiley, 1961.

Dickerson, R. C., & Murphy, B. N. Scope of set scrubbers for odor control. *Annals of the New York Academy of Sciences*, 1974, *237*, 374–388.

Dominic, C. J. Pheromonal mechanisms regulating mammalian reproduction. *General and Comparative Endocrinology*, Suppl. 2, 1969, 269–276.

Doty, R. L. Reproductive endocrine influences upon nasal chemoreception: A review. In R. L. Doty (Ed.), *Mammalian olfaction, reproductive processes, and behavior*. New York: Academic Press, 1976. Pp. 295–321.

Doty, R. L. Food preference ratings of congenitally anosmic humans. In M. R. Kare & O. Maller (Eds.), *The chemical senses and nutrition*. New York: Academic Press, 1977. Pp. 315–325. (a)

Doty, R. L. A review of recent psychophysical studies examining the possibility of communication of sex and reproductive state in humans. In D. Müller-Schwarze & M. M. Mozell (Eds.), *Chemical signals in vertebrates*. New York: Plenum Press, 1977. Pp. 273–286. (b)

Doty, R. L. A review of olfactory dysfunctions in man. *American Journal of Otolaryngology*, 1979, *1*, 57–79.

Doty, R. L., Brugger, W. E., Jurs, P. C., Orndorff, M. A., Snyder, P. J., & Lowry, L. D. Intranasal stimulation from odorous volatiles. Psychometric responses from anosmic and normal humans. *Physiology and Behavior*, 1978, *20*, 175–185.

Doty, R. L., Ford, M., Preti, G., & Huggins, G. R. Changes in the intensity and pleasantness of human vaginal odors during the menstrual cycle. *Science*, 1975, *190*, 1316–1318.

Doty, R. L., Snyder, P. J., Huggins, G. R., & Lowry, L. D. Endocrine, cardiovascular, and psychological correlates of olfactory sensitivity changes during the human menstrual cycle. *Journal of Comparative and Physiological Psychology*, 1981, *95*, 45–60.

Douek, E. *The sense of smell and its abnormalities*. Edinburgh and London: Churchill Livingstone, 1974.

Døving, K. B. Studies of the relation between the frog's electro-olfactogram (EOG) and single unit activity in the olfactory bulb. *Acta Physiologica Scandinavica*, 1964, *60*, 150–163.

Døving, K. B. An electrophysiological study of odour similarities of homologous substances. *Journal of Physiology*, 1966, *186*, 97–109.

References

Døving, K. B. Experiments in olfaction. In G. E. Wolstenhokme & J. Knight (Eds.), *Ciba Foundation symposium on taste and smell in vertebrates.* London: J. and A. Churchill, 1970.

Døving, K. B., & Lange, A. L. Comparative studies of sensory relatedness of odours. *Scandinavian Journal of Psychology,* 1967, *8,* 47–51.

Døving, K. B., & Schieldrop, B. An apparatus based on turbulent mixing for delivery of odorous stimuli. *Chemical Senses and Flavor,* 1975, *1,* 371–374.

Døving, K. B., & Selset, R. Behavior patterns in cod released by electrical stimulation of olfactory tract bundles. *Science,* 1980, *207,* 559–560.

Drake, B., Johansson, B., von Sydow, D., & Døving, K. B. Quantitative psychophysical and electrophysiological data on some odorous compounds. *Scandinavian Journal of Psychology,* 1969, *10,* 89–96.

Dravnieks, A. Approaches to objective olfactometry. In N. Tanyolac (Ed.), *Theories of odors and odor measurement.* Bebek, Istanbul: Robert College Research Center, 1968. Pp. 33–46.

Dravnieks, A., Bock, F. C., Powers, J. J., Tibbetts, M., & Ford, M. Comparison of odors directly and through profiling. *Chemical Senses and Flavor,* 1978, *3,* 191–225.

Dröscher, V. B. *The magic of the senses: New discoveries in animal perception.* New York: Dutton 1969. Pp. 97–145.

Duclaux, R., & Cabanac, M. Psychophysiologie—Effects d'un ingestion de glucose sur la sensation et la perception d'un stimulus olfactif alimentaire. *Comptes Rendus Academie des Sciences* Series D, 1970, *270,* 1006–1009.

Duffee, R., Jann, P. R., Flesh, R. D., & Cain, W. S. *Odor/ventilation relationships in public buildings.* Paper presented at the Seventy-third annual meeting of the Air Pollution Control Association, Montreal, Quebec, June 22–27, 1980. 24 pp.

Duncan, R. B., & Briggs, M. Treatment of uncomplicated anosmia by vitamin A. *Archives of Otolaryngology,* 1962, *75,* 116–124.

Edwards, R. B. Do pleasures and pains differ qualitatively? *Journal of Value Inquiry,* 1975, *9,* 270–281.

Eich, J. E. Fragrances as cues for remembering words. *Journal of Verbal Learning and Verbal Behavior,* 1978, *17,* 103–111.

Eisler, H. How prothetic is the continuum of smell? *Scandinavian Journal of Psychology,* 1963, *4,* 29–32.

Ekman, G. Two generalized ratio scaling methods. *Journal of Psychology,* 1958, *45,* 287–295.

Ekman, G. A direct method for multidimensional rating scaling. *Psychometrika,* 1963, *28,* 33–41.

Ekman, G., Berglund, B., Berglund, V., & Lindvall, T. Perceived intensity of odor as a function of time of adaptation. *Scandinavian Journal of Psychology,* 1967, *8,* 177–186.

Ekman, G., & Engen, T. Multidimensional ratio scaling and multidimensional similarity in olfactory perception. *Reports from the Psychological Laboratory, University of Stockholm,* 1962, No. 126.

Ekman, G., Engen, T., Künnapas, T., & Lindman, R. A quantitative principle of qualitative similarity. *Journal of Experimental Psychology,* 1964, *68,* 530–536.

Ellis, H. *Studies in the psychology of sex.* Volume IV. *Sexual selection in man:* I. Touch. II. Smell. III. Hearing. IV. Vision. Philadelphia: Davis, 1928.

Elsberg, C. A. The sense of smell. VIII. Olfactory fatigue. *Bulletin of the Neurological Institute of New York,* 1935–1936, *4,* 479–495.

Elsberg, C. A., & Levy, I. The sense of smell. A new and simple method of quantitative olfactometry. *Bulletin of the Neurological Institute of New York,* 1935, *4,* 5–19.

Engen, T. Effect of practice and instruction on olfactory thresholds. *Perceptual and Motor Skills*, 1960, *10*, 195–198.
Engen, T. Cross-adaptation to the aliphatic alcohols. *American Journal of Psychology*, 1963, *76*, 96–102.
Engen, T. Psychophysical scaling of odor intensity and quality. *Annals of the New York Academy of Sciences*, 1964, *116*, 504–516.
Engen, T. Olfactory psychophysics. In L. M. Beidler (Ed.), *Handbook of sensory physiology*. Vol. IV. *Chemical senses 1: Olfaction*. New York: Springer-Verlag, 1971. Pp. 216–244. (a)
Engen, T. Psychophysics I. Discrimination and detection. In J. W. Kling & L. A. Riggs (Eds.), *Woodworth and Schlosberg's experimental psychology* (3rd ed.). New York: Holt, Rinehart, and Winston, 1971. Pp. 11–46. (b)
Engen, T. Psychophysics II. Scaling methods. In J. W. Kling & L. A. Riggs (Eds.), *Woodworth and Schlosberg's experimental psychology* (3rd ed.). New York: Holt, Rinehart, and Winston, 1971. Pp. 47–86. (c)
Engen, T. The effect of expectation on judgments of odor. *Acta Psychologica*, 1972, *36*, 450–458. (a)
Engen, T. Use of sense of smell in determining environmental quality. In W. A. Thomas (Ed.), *Indicators of environmental quality*. New York: Plenum Press, 1972. (b)
Engen, T. Method and theory in the study of odor preference. In J. W. Johnston, Jr., D. G. Moulton, & A. Turk (Eds.), *Human responses to environmental odors*. New York: Academic Press, 1974.
Engen, T. Taste and smell. In J. E. Birren & K. Warner Schaie (Eds.), *Handbook of the psychology of aging*. New York: Van Nostrand Reinhold, 1977.
Engen, T. Controlling food preferences in children. In F. D. Horowitz (Ed.), *Early developmental hazards: Predictors and precautions*. New York: Westview Press, 1978. (a)
Engen, T. The origin of preferences in taste and smell. In J. H. A. Kroeze (Ed.), *Preference behaviour and chemoreception*. London: Information Retrieval Limited, 1978. (b)
Engen, T. Why the aroma lingers on. *Psychology Today*, 1980, *13*, 138.
Engen, T., & Berson, D. *Olfactory recognition in the deaf*. Unpublished manuscript, Brown University, 1975.
Engen, T., & Bosack, T. N. Facilitation in olfactory detection. *Journal of Comparative and Physiological Psychology*, 1969, *68*, 320–366.
Engen, T., & Eaton, J. *Free recall of odor and color names*. Unpublished manuscript, Brown University, 1975.
Engen, T., & Furth, P. A. *Effects of cigarette smoking on odor detection*. Unpublished manuscript, Brown University, 1975.
Engen, T., Kilduff, R. A., & Rummo, N. J. The influence of alcohol on odor detection. *Chemical Senses and Flavor*, 1975, *1*, 323–329.
Engen, T., Kuisma, J. E., & Eimas, P. D. Short-term memory of odors. *Journal of Experimental Psychology*, 1973, *99*, 222–225.
Engen, T., & Lindström, C. O. Psychophysical scales of the odor intensity of amyl acetate. *Scandinavian Journal of Psychology*, 1963, *4*, 23–28.
Engen, T., & Lipsitt, L. P. Decrement and recovery of responses to olfactory stimuli in the human neonate. *Journal of Comparative and Physiological Psychology*, 1965, *59*, 312–316.
Engen, T., Lipsitt, L. P., & Kaye, H. Olfactory responses and adaptation in the human neonate. *Journal of Comparative and Physiological Psychology*, 1963, *56*, 73–77.
Engen, T., & McBurney, D. H. Magnitude and category scales of the pleasantness of odors. *Journal of Experimental Psychology*. 1964, *68*, 435–440.

Engen, T., & Omark, J. N. *An exploration of autocentric versus allocentric perception with odors and pictures.* Unpublished report, Brown University, 1977.
Engen, T., & Pfaffmann, C. Absolute judgments of odor intensity. *Journal of Experimental Psychology,* 1959, *58,* 23–26.
Engen, T., & Pfaffmann, C. Absolute judgments of odor quality. *Journal of Experimental Psychology,* 1960, *59,* 214–219.
Engen, T., & Ross, B. M. Long-term memory of odors with and without verbal descriptions. *Journal of Experimental Psychology,* 1973, *100,* 221–227.
Erickson, R. P. Stimulus coding in topographic and nontopographic afferent modalities: On the significance of the activity of individual sensory neurons. *Psychological Review,* 1968, *75,* 447–465.
Erickson, R. P., & Schiffmann, S. S. The chemical senses: A systematic approach. *Handbook of psychobiology.* New York: Academic Press, 1975. Pp. 393–426.
Eyferth, K. Über die Abhängigkeit der Urteile über olfactorische Qualitäten von der Emphindung der Reizintensität. *Zeitschrift für Experimentale und Angewandte Psychologie,* 1965, *12,* 209–222.
Eyferth, K., & Krueger, K. The construction of a programmable olfactometer. *Behavioral Research Methodology and Instrumentation.* 1970, *2,* 19–22.
Fazzalari, F. A. (Ed.). *Compilation of odor and taste threshold values data.* Philadelphia: American Society for Testing and Materials, 1978.
Fechner, G. T. *Elemente der Psychophysik.* Leipzig: Breitkopf and Harterl, 1860. (English translation of Vol. 1 by H. E. Adler. Holt, Rinehart and Winston, 1966.)
Fledstein, M., Levaggi, D. A., & Thuillier, R. Odor regulation by emission limitation at the stack. *Annals of the New York Academy of Sciences,* 1974, *237,* 309–314.
Flesh, R. D. Social and economic criteria for odor control effectiveness. *Annals of the New York Academy of Sciences,* 1974, *237,* 320–327.
Foreyt, J. P., & Kennedy, W. A. Treatment of overweight by aversion therapy. *Behavior Research and Therapy,* 1971, *9,* 29–34.
Foster, D., Scofield, E. H., & Dallenbach, K. M. An olfactorium. *American Journal of Psychology,* 1950, *63,* 431–440.
Free, E. E. Shall we train our noses? *Forum,* 1926, *75,* 45–50.
Freud, S. *Civilization and its discontents.* (J. Stachey, trans.) New York: Norton, 1962.
Fribers, L., Jonsson, E., & Cederlöf, R. Studies of hygienic nuisances of waste gases from sulfur pulp mill. Part I. An interview investigation. *Nordisk Hygienisk Tidskrift,* 1960, *41,* 41–50. (In Swedish) (a)
Friberg, L., Jonsson, E., & Cederlöf, R. Studies of hygienic nuisances of waste gases from sulfur pulp mill. Part II. Odor threshold determinations. *Nordisk Hygienisk Tidskrift,* 1960, *41,* 50–62. (In Swedish) (b)
Frijters, J. E. R., & Beumer, S. C. C. Psychophysical methodology in odour pollution research: The measurement of poultry house odour detectability and intensity. *Chemical Senses and Flavor,* 1979, *4,* 327–340.
Fukomoto, Y., Nakajima, H., Vetake, M., Masuyama, A., & Yoshida, T. Smell ability to solutions of potassium cyanide and its inheritance. *Japan Journal of Human Genetics,* 1957, *2,* 7–16. (In Japanese)
Furchtgott, E., & Friedman, M. P. The effect of hunger and taste on odor RLs. *Journal of Comparative and Physiological Psychology,* 1960, *53,* 576–581.
Gamble, E. M. The applicability of Weber's law to smell. *American Journal of Psychology,* 1898, *10,* 82–142.
Garcia, J., & Brett, L. P. Conditioned responses to food odor and taste in rats and wild

predators. In M. R. Kare & O. Maller (Eds.), *The chemical senses and nutrition.* New York: Academic Press, 1977. Pp. 277–290.
Garcia-Medina, M. R. Flavor–odor taste interactions in solutions of acetic acid and coffee. *Chemical Senses,* 1981, *6,* 13–22.
Garner, W. R. The stimulus in information processing. *American Psychologist,* 1970, *25,* 350–358.
Gesteland, R. C. The neural code: Integrative neural mechanisms. In E. C. Carterette & M. P. Friedman (Eds.), *Handbook of perception.* Vol. VIA. *Tasting and smelling.* New York: Academic Press, 1978. Pp. 259–276.
Gesteland, R. C., Lettvin, J. Y., & Pitts, H. W. Chemical transmission in the nose of the frog. *Journal of Physiology,* 1965, *181,* 525–559.
Gesteland, R. C., Lettvin, J. Y., Pitts, W. H., & Rojas, A. Odor specificities of the frog's olfactory receptors. In Y. Zotterman (Ed.), *Olfaction and taste.* Oxford: Pergammon Press, 1963.
Getchell, T. V. Electrogenic sources of slow voltage transients recorded from frog olfactory epithelium. *Journal of Neurophysiology,* 1974, *37,* 1115–1130.
Gibson, J. J. *The senses considered as perceptual systems.* Boston: Houghton Mifflin, 1966.
Glaser, O. Hereditary deficiencies in the sense of smell. *Science,* 1918, *48,* 647–648.
Gloor, P. Temporal lobe epilepsy: Its possible contribution to the understanding of the functional significance of the amygdala and its interaction with neocortical–temporal mechanisms. In B. E. Eleftheriou (Ed.), *The neurobiology of the amygdala.* New York: Plenum Press, 1972. Pp. 423–457.
Goldfoot, D. A., Kravetz, M. A., Goy, R. W., & Freeman, S. K. Lack of effect of vaginal lavages and aliphatic acids on ejaculatory responses in rhesus monkeys: Behavioral and chemical analyses. *Hormones and Behavior,* 1976, *7,* 1–27.
Gordon, K. The recollection of pleasant and unpleasant odors. *Journal of Experimental Psychology,* 1925, *8,* 225–239.
Graziadei, P. P. C., & Graziadei, G. A. M. The olfactory system: A model for the study of neurogenesis and axon regeneration in animals. In C. W. Cotman (Ed.), *Neuronal plasticity.* New York: Raven Press, 1978. Pp. 131–154.
Green, D. M., & Swets, J. A. *Signal detection theory and psychophysics.* New York: Wiley, 1966. (Reprinted New York: Krieger, 1974.)
Green, P. E., & Rao, V. R. *Applied multidimensional scaling.* New York: Holt, 1972.
Grennfelt, P., & Lindvall, T. Sensory and physical–chemical studies of pulp mill odors. In *Proceedings of the Third International Clean Air Congress.* Düsseldorf VDI-Verlag, 1973. Pp. A36–A39.
Gregson, R. A. M., & Mitchell, M. J. Odor quality similarity scaling and odor–word profile matching. *Chemical Senses and Flavor,* 1974, *1,* 95–101.
Guillot, M. Anosmies partielles et odeurs fondamentales. *Canadian Royal Academy of Science Service,* 1948, *226,* 1307–1309.
Hagan, P. J. Posttraumatic anosmia. *Archives of Otolaryngology,* 1967, *85,* 107–111.
Hamauzu, Y. Odor perception measurement by the use of odorless room. *Sangyo Kogai* [Industrial Public Nuisance], 1969, *5,* 718–723. (In Japanese)
Hammond, D. J. The head law—How will you vote? *Motor Boating and Sailing,* January 1980, p. 6.
Hardison, L. C., & Steenberg, L. R. Criteria for chemical odor control systems. *Annals of the New York Academy of Sciences,* 1974, *237,* 363–373.
Haring, H. G. Vapor pressures and Raoult's law deviations in relation to odor enhancement and suppression. In A. Turk, J. W. Johnston, Jr., & D. G. Moulton (Eds.), *Human responses to environmental odors.* New York: Academic Press, 1974. Pp. 199–226.

References

Haring, H. G., Turk, A., & Okey, R. W. Cost-effectiveness relationships in odor control. *Annals of the New York Academy of Sciences,* 1974, *237,* 328–338.

Harper, R., Bate-Smith, E. C., & Land, D. G. *Odour description and odour classification.* London: J. & A. Churchill, 1968.

Hazzard, F. W. A descriptive account of odors. *Journal of Experimental Psychology,* 1930, *13,* 297–331.

Hellman, T. M., & Taylor, B. R. Odor control by high-temperature oxidation. *Annals of the New York Academy of Sciences,* 1974, *237,* 350–362.

Helson, H. *Adaptation-level theory.* New York: Harper & Row, 1974.

Henion, K. E. Pleasantness and intensity: A single dimension? *Journal of Experimental Psychology,* 1971, *90,* 275–279.

Henkin, R. I. Hypogonadism associated with familial hyposmia. *Clinical Research,* 1965, *13,* 244.

Henkin, R. I., & Bartter, F. C. Studies on olfactory thresholds in normal man and in patients with adrenal cortical steroids and of serum sodium concentration. *Journal of Clinical Investigation,* 1966, *45,* 1631–1639.

Henkin, R. I., Hoye, R. C., Ketcham, A. S., & Gould, W. J. Hyposmia following laryngectomy. *Lancet,* 1968, *2,* 479–481.

Henkin, R. I., & Powell, G. F. Increased sensitivity of taste and smell in cystic fibrosis. *Science,* 1962, *138,* 1107–1108.

Henkin, R. I., Schechter, P. J., Hoye, R., & Mattern, C. F. T. Ideopathic hypogeusia with dysgeusia, hyposmia, and dysosmia. A new syndrome. *American Medical Assocation Journal,* 1971, *217,* 434–440.

Henkin, R. I., & Powell, G. F. Appetite and anosmia. *Lancet,* 1971, *1,* 1352–1353.

Henning, H. *Der Geruch* (rev. ed.) Leipzig: Barth, 1916. (Revised 1924.)

Herriot, J. *All things wise and wonderful.* New York: St. Martin's Press, 1977.

Hertz, J., Cain, W. S., Bartoshuk, L., & Dolan, T. F., Jr. Olfactory and taste sensitivity in children with cystic fibrosis. *Physiology and Behavior,* 1975, *14,* 89–94.

Hold, B., & Schleidt, M. The importance of human odour in non-verbal communication. *Zeitschrift Tierpsychologic,* 1977, *43,* 225–238.

Hopkins, H. The smell that tells. *FDA Consumer,* December 1974–January 1975, DHEW Publications No. (FDA) 75-2022.

Horstman, S. W., Wromble, R. F., & Heller, A. N. Identification of community odor problems by use of an observer corps. *Journal of the Air Pollution Control Association,* 1971, *21,* 341–346.

Howard, J. Mr. Bellow considers his planet. *Life,* 1970, *68,* (April 3), 57–58.

Hubert, H. B., Fabsitz, R. R., Feinleib, M., & Brown, K. S. Olfactory sensitivity in humans: Genetic versus environmental control. *Science,* 1980, *208,* 607–609.

Hughes, J. R., & Mazurowski, J. A. Studies on the supracallosal cortex in unanesthetized mammals. II. Monkeys. B. Responses from olfactory bulb. *Electroencephalography and Clinical Neurophysiology,* 1962, *14,* 635–645.

Jacobson, M., Beroza, M., & Jones, W. A. Isolation, Identification, and synthesis of the sex attractant of gypsy moth. *Science,* 1960, *132,* 1011.

James, W. *Psychology.* New York: Holt, 1893.

Jamieson, D. G. Visual influence on taste sensitivity. *Perception and Psychophysics,* in press.

Johnston, J. W., Jr. Quantification of olfactory stimuli. In C. C. Brown (Ed.), *Methods of psychophysiology.* Baltimore: Williams and Wilkins, 1967. Pp. 192–220.

Johnston, J. W., Jr. The essential nature of the putrid odors. In N. Tanyolac (Ed.), *Theories of odors and odor measurement.* Bebek, Istanbul: Robert College Research Center, 1968.

Jones, F. N. A test of the validity of the Elsberg method of olfactometry. *American Journal of Psychology,* 1953, *46,* 81–85.

Jones, F. N. Olfactory absolute thresholds and their implications for the nature of the receptor process. *Journal of Psychology,* 1955, *40,* 223–227.
Jones, F. N. Information content of olfactory quality. In N. Tanyoloac (Ed.), *Theories of odors and odor measurement.* Bebek, Istanbul: Robert College Research Center, 1968.
Jones, F. N., & Jones, M. H. Modern theories of olfaction: A critical review. *Journal of Psychology,* 1953, *36,* 207–241.
Jones, F. N., Roberts, K., & Holman, E. W. Similarity judgments and recognition memory for some common spices. *Perception and Psychophysics,* 1978, *24,* 2–6.
Jones, F. N., & Woskow, M. H. On the intensity of odor mixtures. *Annals of the New York Academy of Sciences,* 1964, *116,* 484–494.
Jones, M. H. A study of the "common chemical sense." *American Journal of Psychology,* 1954, *67,* 696–699.
Kaiser, E. R. Odor and its measurement. In A. C. Stern (Ed.), *Air pollution.* Vol. 1. New York: Academic Press, 1963.
Kaissling, K. E. Insect olfaction. In L. M. Beidler (Ed.), *Handbook of sensory physiology,* Volume IV. *Chemical senses 1: Olfaction.* New York: Springer-Verlag, 1971. Pp. 351–431.
Kallmann, F. J., Schoenfeld, W. A., & Barrera, S. E. The genetic aspects of primary eunuchoidism. *American Jounrnal of Mental Deficiency,* 1944, *48,* 203–236.
Kalmus, H. The discrimination by the nose of the dog of individual human odours and in particular the odour of twins. *British Journal of Animal Behavior,* 1955, *3,* 25–31.
Kalogerakis, M. G. The role of olfaction in sexual development. *Psychosomatic Medicine,* 1963, *25,* 420–432.
Kaplan, J. N., Cubicciotti, D. D., III, & Redican, W. K. Olfactory and visual differentiation of synthetically scented surrogates by infant squirrel monkeys. *Developmental Psychobiology,* 1979, *12,* 3–19.
Kendall, D. A., & Neilsen, A. J. Sensory and chromatographic analysis of mixtures formulated from pure odorants. *Journal of Food Science,* 1966, 1960, *31,* 268–274.
Kennedy, W. A., & Forety, J. P. Control of eating behavior in an obese patient by avoidance conditioning. *Psychological Reports,* 1968, *22,* 571–576.
Kenneth, J. H. *Osmics: The science of smell.* Edinburgh: Oliver and Boyd, 1922.
Kern, S. Olfactory ontology and scented harmonies: On the history of smell. *Journal of Popular Culture,* Spring, 1974, 814–824.
Keverne, E. B. Sex attractants in primates. *Journal of the Society of Cosmetic Chemists,* 1976, *27,* 257–269.
Kimbrell, G. McH., & Furchgott, E. The effect of aging on olfactory threshold. *Journal of Gerontology,* 1963, *18,* 364–365.
Kirk, R. L., & Stenhause, N. S. Ability to smell solutions of potassium cyanide. *Nature,* 1953, *171,* 698–699.
Kirk-Smith, M., & Booth, D. Effect of androstenone on choice of location in others' presence. In H. van der Starre (Ed.), *Olfaction and taste VII.* London: IRL Press, 1980. Pp. 397–400.
Kirk-Smith, M., Booth, D., Carrol, D., & Davies, P. Human sexual attitudes affected by androstenol. *Research Communications in Psychology, Psychiatry, and Behavior,* 1978, *3,* 379–384.
Kniep, H. H., Morgan, W. L., & Young, P. T. Studies in affective psychology, XI. Individual differences in affective reactions to odors. *American Journal of Psychology,* 1931, *43,* 406–421.
Koelega, H. S. Olfaction and sensory asymmetry. *Chemical Senses and Flavor,* 1979, *4,* 89–95.
Koelega, H. S., & Köster, E. P. Some experiments on sex differences in odor perception. *Annals of the New York Academy of Sciences,* 1974, *237,* 234–246.

Köster, E. P. Olfactory sensitivity and the menstrual cycle. *International Rhinology,* 1965, *3,* 57–64.
Köster, E. P. Relative intensity of odour mixtures at supra-threshold level. *Olfactologia. Cashier's D'oto-Rhino' Laryngologie,* 1968, III, Suppl. 5.
Köster, E. P. Intensity in mixtures of odorous substances. In C. Pfaffmann (Ed.), *Olfaction and taste, III.* New York: Rockefeller University Press, 1969. Pp. 142–149.
Köster, E. P. *Adaptation and cross-adaptation in olfaction.* Ph.D. thesis, University of Utrecht, Netherlands. Rotterdam/New York: Bronder-Offsett, 1971.
Köster, E. P., & Macleod, P. Psychophysical electrophysiological experiments with binary mixtures of acetophenone and eugenol. In D. G. Moulton, A. Turk, & J. W. Johnston, Jr. (Eds.), *Methods in olfactory research.* New York: Academic Press, 1975. Pp. 431–444.
Krauskopf, J. Effect of retinal image stabilization in the appearance of heterochromatic targets. *Journal of the Optical Society of America,* 1963, *53,* 741–743.
Kunst-Wilson, W. R., & Zajonc, R. B. Affective discrimination of stimuli that cannot be recognized. *Science,* 1980, *207,* 557–558.
Laffort, P. A linear relationship between olfactory effectiveness and identified molecular characteristics, extended to fifty pure substances. In C. Pfaffmann (Ed.), *Olfaction and taste, III.* New York: Rockefeller University Press, 1969. Pp. 150–157.
Laffort, P., & Dravnieks, A. An approach to a physico-chemical model of olfactory stimulation in vertebrates by single compounds. *Journal of Theoretical Biology,* 1973, *38,* 335–345.
Laing, D. G. A comparative study of olfactory sensitivity of humans and rats. *Chemical Senses and Flavor,* 1975, *1,* 257–269.
Laing, D. G., & MacKay-Sim, A. Olfactory adaptation in the rat. In D. A. Denton & J. P. Coghlan (Eds.), *Olfaction and taste, V.* New York: Academic Press, 1975. Pp. 291–295.
Laing, D. G., Panhuber, H., & Baxter, R. I. Olfactory properties of amines and *n*-butanol. *Chemical senses and flavor,* 1978, *3,* 149–166.
Land, D. G. Hedonic responses and perceived characteristics of odours in man. In J. H. A. Kroeze (Ed.), *Preference behavior and chemoreception.* London: Information Retrieval Limited, 1979. Pp. 93–106.
Lange, A. *Multivariata studier i luitperception.* Unpublished thesis, University of Stockholm, 1970.
Lashley, K. S., & Sperry, R. W. Olfactory discrimination after destruction of the anterior thalamic nuclei. *American Journal of Physiology,* 1943, *139,* 446–450.
Lawless, H. T. The pleasantness of mixture in taste and olfaction. *Sensory Processes,* 1977, *1,* 227–237.
Lawless, H. T. Recognition of common odors, pictures, and simple shapes. *Perception and Psychophysics,* 1978, *24,* 493–495.
Lawless, H. T., & Cain, W. S. Recognition memory for odors. *Chemical Senses and Flavor,* 1975, *1,* 331–337.
Lawless, H. T., & Engen, T. Associations to odors: Interference, memories, and verbal labeling. *Journal of Experimental Psychology,* 1977, *3,* 52–59. (a)
Lawless, H. T., & Engen, T. Memory. *McGraw-Hill yearbook of science and technology.* New York: McGraw-Hill, 1977. Pp. 299–301. (b)
Leigh, A. D. Defects of smell after head injury. *Lancet,* 1943, *1,* 38–40
LeMagnen, J. Analyses d'odours complexes et homologues par fatigue. *Comptes Rendus Academie des sciences.* 1948, *226,* 753–754.
LeMagnen, J. Nouvelles donnés sur le hénomène de l'exaltolide. *Comptes Rendus Academie des Sciences,* 1950, *230,* 1103.
LeMagnen, J. Les phénomènes olfacto-sexuels chez l'homme. *Archives of Science and Physics,* 1952, *6,* 125–160.

LeMagnen, J. Olfaction and nutrition. In L. M. Beidler (Ed.), *Handbook of sensory physiology.* Vol. IV. *Chemical senses 1: Olfaction.* New York: Springer-Verlag, 1971. Pp. 465–482.

Lettvin, J. Y., & Gesteland, R. C. Speculations on smell. *Cold Spring Harbor Symposium on Quantitative Biology,* 1965, *30,* 217–225.

Leveteau, J., & Macleod, P. Olfactory discrimination in the rabbit olfactory glomerulus. *Science,* 1966, *153,* 175–176.

Levine, J. M., & McBurney, D. H. Causes and consequences of effluvia: Body odor awareness and controllability as determinants of interpersonal evaluation. *Personality and Social Psychology Bulletin,* 1977, *3,* 442–445.

Lewis, W. L., & Executors of C. S. Lewis. *The letters of C. S. Lewis.* New York: Harcourt Brace Jovanovich, 1966.

Lindvall, T. Measurement of odorous air pollutants. *Nordisk Hygienisk Tidskrift,* 1966, *47,* 41–71. (In Swedish)

Lindvall, T. Nuisance effects of air pollutants. *Nordisk Hygienisk Tidskrift,* 1969, *50,* 99–114.

Lindvall, T. On sensory evaluation of odorous air pollutant intensities. Measurements of odor intensity in the laboratory and in the field with special reference to effluents of sulfate pulp factories. *Nordisk Hygienisk Tidskrift,* 1970, Supplement 2, 1–182.

Lindvall, T. Sensory measurement of ambient traffic odors. *Journal of Air Pollution Control Association,* 1973, *23,* 697–700.

Lindvall, T. Monitoring odorous air pollution in the field with human observers. *Annals of the New York Academy of Sciences,* 1974, *237,* 247–260.

Lindvall, T., Norén, O., & Thyselius, L. On the abatement of animal manure odours. In *Proceedings of the Third International Clean Air Congress.* Dusseldorf: VDI-Verlag, 1973. Pp. E120–E123.

Lindvall, T., & Radford, E. P. (Eds.). Measurement of annoyance due to exposure to environmental factors. *Environmental Research,* 1973, *6,* 1–36.

Lindvall, T., & Svensson, L. T. Equal unpleasantness matching of malodorous substances in the community. *Journal of Applied Psychology,* 1974, *59,* 264–269.

Lipsitt, L. P., Engen, T., Bloom, S. J., & Jennings. V. Olfactory development in children to five years of age. Unpublished report, Brown University, 1975.

Lipsitt, L. P., Engen, T., & Kaye, H. Developmental changes in the olfactory threshold of the neonate. *Child Development,* 1963, *34,* 371–376.

London, I. D. Research on sensory interaction in the Soviet Union. *Psychological Bulletin,* 1954, *51,* 531–568.

Loo, S. K. A comparative study of the histology of the nasal fossa in four primates. *Folia Primatologica,* 1973, *20,* 410–422.

Macfarlane, A. Olfaction in the development of social preferences in the human neonate. *Ciba Foundation Symposium,* 1975, *33,* (New Series), 103–117.

MacLeod, P. Structure of higher olfactory centers. In L. M. Beidler (Ed.), *Handbook of sensory physiology.* Vol. IV. *Chemical senses 1: Olfaction.* New York: Springer-Verlag, 1971.

Mair, R. G., Bouffard, J. A., Engen, T., & Morton, T. H. Olfactory sensitivity during the menstrual cycle. *Sensory Processes,* 1978, *2,* 90–98.

Mair, R. G., Capra, C., McEntee, W. J., & Engen, T. Odor discrimination and memory in Korsakoff's psychosis. *Journal of Experimental Psychology: Human Perception and Performance,* 1980, *6,* 445–458.

Mair, R. G., & Engen, T. Some effects of aphasic lesions on odor perception. *Sensory Processes,* 1976, *1,* 33–39.

Males, J. L., Townsend, J. L., & Schneider, R. A. Hypogonadotrophic hypogonadism with anosmia. *Archives of Internal Medicine,* 1973, *131,* 501–507.

Marshall, D. A., & Moulton, D. G. Olfactory sensitivity to α-ionine in humans and dogs. *Chemical Senses,* 1981, *6,* 53–61.

Martin, S., & Pangborn, R. M. A note on responses to ethyl alcohol before and after smoking. *Perception and Psychophysics,* 1970, *8,* 169–170.
Matthews, W. B., & Rundle, A. T. Familial cerebral ataxia and hypogonadism. *Brain,* 1964, *87,* 463–468.
McBurney, D. H. Effects of adaptation on human taste function. In C. Pfaffmann (Ed.), *Olfaction and taste, III.* New York: Rockefeller University Press, 1969. Pp. 407–419.
McBurney, D. H., Levine, J. M., & Cavanaugh, P. H. Psychophysical and social ratings of human body odor. *Personality and Social Psychology Bulletin,* 1977, *3,* 135–138.
McBurney, D. H., & Moskat, L. J. Taste thresholds in college-age smokers and nonsmokers. *Perception and Psychophysics,* 1975, *18,* 71–73.
McCartney, W. *Olfaction and odours.* New York: Springer-Verlag, 1968.
McClintock, M. K. Menstrual synchrony and suppression. *Nature,* 1971, *229,* 224–225.
Mesolella, V. L'Olfatto nelle diverse eta. *Archivo Italiano di Otologia Rinologia e Laringologia,* 1943, *46,* 43–62.
Michael, R. P., Bonsall, R. W., & Warner, P. Human vaginal secretions: Volatile fatty acid content. *Science,* 1974, *186,* 1217–1219.
Michael, R. P., & Keverne, E. B. Pheromones in the communication of sexual status in primates. *Nature,* 1968, *218,* 746–749.
Michael, R. P., Keverne, E. B., & Bonsall, R. W. Pheromones: Isolation of male sex attractants from a female primate. *Science,* 1971, *172,* 964–965.
Millen, J. K. *Your nose knows.* Los Angeles: Cunningham Press, 1960.
Miller, G. A. The magican number seven, plus or minus two: Some limitations on our capacity for processing information. *Psychological Review,* 1956, *63,* 81–97.
Miner, S. *Preliminary air pollution survey of hydrogen sulfide. A literature review.* National Air Pollution Control Administration Publication No. APTD 69-37. Bethesda, Md.: Litton Systems, 1969. (Available from National Technical Information Service as Publication No. PB-188 068.)
Mitchell, M. A., Konigsbacher, K. S., & Edman, W. M. The importance of odor as a nonfunctional component or Odor—a tool of marketing. *Annals of the New York Academy of Sciences,* 1964, *116,* 685–691.
Mitchell, M. J., Gregson, R. A. M. Interrelations of perceptual reports of smell, taste and irritance over the near-threshold range. *Perception and Psychophysics,* 1968, *4,* 13–18.
Moncrieff, R. W. What is odor? A new theory. *Essential Oil Review,* 1949, *54,* 453–454.
Montcrieff, R. W. Olfactory adaptation and odor likeness. *Journal of Physiology,* (London), 1956, *133,* 301–316.
Montcrieff, R. W. *Odour preferences.* New York: Wiley, 1966.
Montcrieff, R. W. *The chemical senses* (3rd ed.). London: Leonard Hill, 1967.
Montcrieff, R. W. Sensory discrimination and smoking. *Food Processes and Marketing,* 1968, *37,* 303–305.
Morris, N. M., & Udry, J. R. Pheromonal inferences on human sexual behaviour: An experimental search. *Journal of Biosocial Science,* 1978, *10,* 147–157.
Moskowitz, H. R. Intensity and hedonic functions for chemosensory stimuli. In M. R. Kare & O. Maller (Eds.), *The chemical senses and nutrition.* New York: Academic Press, 1977. Pp. 71–101.
Moskowitz, H. R. Taste and food technology: Acceptability, aesthetics, and preference. In E. C. Carterette & M. P. Friedman (Eds.), *Handbook of perception.* Vol. VIA. *Tasting and smelling.* New York: Academic Press, 1978.
Moskowitz, H. R. Mind, body and pleasure: An analysis of factors which influence sensory hedonics. In J. H. A. Kroeze (Ed.), *Preference behaviour and chemoreception.* London: Information Retrieval Limited, 1979. Pp. 131–147. (a)

Moskowitz, H. R. Utility of the vector for higher-order mixtures: A correction. *Sensory Processes*, 1979, *3*, 366–369. (b)

Moskowitz, H. R., & Barbe, C. D. Profiling of odor components and their mixtures. *Sensory Processes*, 1977, *1*, 212–226.

Moskowitz, H. R., Dravnieks, A., Cain, W. S., & Turk, A. Standardized procedure for expressing odor intensity. *Chemical Senses and Flavor*, 1974, *1*, 235–237.

Moskowitz, H. R., Dravnieks, A., & Klarman, L. A. Odor intensity and pleasantness for a diverse set of odorants. *Perception and Psychophysics*, 1976, *19*, 122–128.

Moulton, D. G. The olfactory pigment. In L. M. Beidler (Ed.) *Handbook of sensory physiology* (Vol. IV): *Chemical senses*. Part 1. *Olfaction*. New York: Springer Verlag, 1971. pp. 59–74.

Moulton, D. G. Minimum odorant concentrations detectable by the dog and their implications for olfactory receptor sensitivity. In D. Müller-Schwarze & M. M. Mozell (Eds.), *Chemical signals in vertebrates*. New York: Plenum Press, 1976. Pp. 455–464. (a)

Moulton, D. G. Spatial patterning of response to odors in the peripheral olfactory system. *Physiological Reviews*, 1976, *56*, 578–593. (b)

Moulton, D. G., Ashton, E. H., & Eayrs, J. T. Studies in olfactory acuity. 4. Relative detectability of *n*-aliphatic acids by the dog. *Animal Behaviour*, 1960, *8*, 117–128.

Moulton, D. G., & Beidler, L. M. Structure and function in the peripheral olfactory system. *Physiological Review*, 1967, *47*, 1–52.

Mower, G. D., Mair, R. G., & Engen, T. Influence of internal factors on the perceived intensity and pleasantness of gustatory and olfactory stimuli. In M. R. Kare & O. Maller (Eds.), *The chemical senses and nutrition*. New York: Academic Press, 1977.

Mozell, M. M. The spatiotemporal analysis of odorants at the level of the olfactory receptor sheet. *Journal of General Physiology*, 1966, *50*, 25–41.

Mozell, M. M. Evidence for a chromatographic model of olfaction. *Journal of General Physiology*, 1970, *56*, 46–63.

Mozell, M. M. The chemical senses. II. Olfaction. In J. W. Kling & L. A. Riggs (Eds.), *Woodworth and Schlosberg's experimental psychology* (3rd ed.). New York: Holt, Rinehart, and Winston, 1971. Pp. 193–222.

Mozell, M. M., & Jagodowicz, M. Chromatographic separation of odorants by the nose: Retention times measured across in vivo olfactory mucosa. *Science*, 1973, *181*, 1247–1249.

Mozell, M. M., Smith, B. P., Smith, P. E., Sullivan, R. J., Jr., & Swender, P. Nasal chemoreception in flavor identification. *Archives of Otolaryngology*, 1969, *90*, 131–137.

Mullins, L. J. Olfaction. *Annals of the New York Academy of Sciences*, 1955, *62*, 247–276.

Murphy, C. The effect of age on taste sensitivity. In S. S. Han & D. H. Coons (Eds.), *Symposium on biology of special senses in aging*. Ann Arbor: University of Michigan Institute of Gerontology, 1979. Pp. 21–33.

Murphy, C., & Cain, W. S. Taste and olfaction: Independence versus interaction. *Physiology and Behavior*, 1980, *24*, 601–605.

Murphy, C., Cain, W. S., & Bartoshuk, L. M. Mutual action of taste and olfaction. *Sensory Processes*, 1977, *1*, 204–211.

Mustaparta, H. Spatial distribution of receptor responses to stimulation with different odours. *Acta Physiologica Scandinavica*, 1971, *82*, 154–166.

Nabokov, V. *Mary*. (Michael Glenny, trans.) New York: McGraw-Hill, 1970.

Nicoll, R. A. Recurrent excitation of secondary olfactory neutrons: A possible mechanism for signal amplification. *Science*, 1971, *171*, 824–826.

O'Connell, R. J. Olfaction, pheromones and the major histocompatibility complex. *Federation Proceedings*, 1978, *37*, 2099–2101.

O'Mahony, M. O. Smell illusion and suggestion: Reports of smells contingent on tones played on television and radio. *Chemical Senses and Flavor*, 1978, *3*, 183–189.

Nyby, J., Whitney, G., Schmitz, S., & Dizinno, G. Postpubertal experience establishes signal value of mammalian sex odor. *Behavioral Biology,* 1978, *22,* 545–552.
Ottoson, D. Analysis of the electrical activity of the olfactory epithelium. *Acta Physiologica Scandinavica,* 1956, *35,* Suppl. 122, 1–83.
Ottoson, D. The electro-olfactogram. In L. M. Beidler (Ed.), *Handbook of sensory physiology.* Vol. IV. *Chemical senses 1: Olfaction.* New York: Springer-Verlag, 1971.
Ozbaydar, S. The effects of darkness and light on auditory sensitivity. *British Journal of Psychology,* 1961, *52,* 285–291.
Pager, J. Nutritional states, food odors, and olfactory function. In M. R. Kare & O. Maller (Eds.), *The chemical senses and nutrition.* New York: Academic Press, 1977. Pp. 51–68.
Pager, J., & Royet, J. P. Some effects of conditioned aversion on food intake and olfactory bulb electrical responses in the rat. *Journal of Comparative and Physiological Psychology,* 1976, *90,* 67–77.
Palmerino, C., Rusiniak, K. W., & Garcia, J. Flavor–illness aversions: The peculiar roles of odor and taste in memory for poison. *Science,* 1980, *208,* 753–755.
Pangborn, R. M. Flavor perception. Relation of sensory to instrumental measurements. In D. J. Tilgner & A. Borys (Eds.), *Proceedings of the Second International Congress of Food Science and Technology.* Warsaw, Poland: Wydawnictwo Przemslu Lekkiego i Spozywczego, 1966.
Pangborn, R. M., Berg, H. W., Roessler, E. B., & Webb, A. D. Influence of methodology on olfactory response. *Perceptual and Motor Skills,* 1964, *18,* 91–103.
Pangborn, R. M., Trabue, I. M., & Barylko-Pikielna, N. Taste, odor, and tactile discrimination before and after smoking. *Perception and Psychophysics,* 1967, *2,* 529–532.
Pantaleoni, R. Critique of communication entitled "The enzyme model of olfaction, and nature of odori-vectors and specific malodor counteractants" by Alfred A. Schleppnik, Monsanto Flavor/Essence. *Perfumer and Flavorist,* 1976, *1,* 16–17.
Parker, G. H. *Smell, taste, and allied senses in the vertebrates.* Philadelphia: Lippincott, 1922.
Patte, F., Etcheto, M., & Laffort, P. Selected and standardized values of suprathreshold odor intensities for 110 substances. *Chemical Senses and Flavor,* 1975, *1,* 283–305.
Patte, F., & Laffort, P. An alternative model of olfactory quantitative interaction in binary mixtures. *Chemical Senses and Flavor,* 1979, *4,* 267–274.
Patterson, P. M., & Lauder, B. A. The incidence and probable inheritance of "smell blindness." *Journal of Heredity,* 1948, *39,* 295–297.
Pelosi, P., & Viti, R. Specific anosmia to *l*-carvone: The minty primary odor. *Chemical Senses and Flavor,* 1978, *3,* 331–337.
Peterson, L. R., & Peterson, M. J. Short-term retention of individual verbal items. *Journal of Experimental Psychology,* 1959, *58,* 193–198.
Peto, E. Contribution to the development of smell feeling. *British Journal of Medical Psychology,* 1936, *15,* 314–320.
Pfaffmann, C. Taste and smell. In S. S. Stevens (Ed.), *Handbook of experimental psychology.* New York: Wiley, 1951.
Pfaffmann, C. The pleasures of sensation. *Psychological Review,* 1960, *67,* 253–268.
Potter, H., & Butters, N. An assessment of olfactory deficits in patients with damage to prefrontal cortex. *Neuropsychologia,* 1980, *18,* 621–628.
Pribram, K. H., & Kruger, L. Functions of the "olfactory brain." *Annals of the New York Academy of Sciences,* 1954, *58,* 109–138.
Principato, J. J., & Ozenberger, J. M. Cyclical changes in nasal resistance. *Archives of Ontolaryngology,* 1970, *91,* 71–77.
Pritchard, R. M. Stabilized images on the retina. *Scientific American,* June 1961, pp. 72–78.
Proust, M. *Swann's Way.* New York: Modern Library, 1928.
Pryor, G. T., Steinmetz, G., & Stone, H. Changes in absolute detection threshold and in

subjective intensity of suprathreshold stimuli during olfactory adaptation and recovery. *Perception and Psychophysics,* 1970, *8,* 331–335.

Putnam, S. *The portable Rabelais.* New York: Viking, 1960. P. 478.

Rausch, R., & Serafetinides, E. A. Specific alterations of olfactory functions in humans with temporal lobe lesions. *Nature,* 1975, *225,* 557–558.

Rausch, R., Serafetinides, E. A., & Crandall, P. H. Olfactory memory in patients with anterior temporal lobectomy. *Cortex,* 1977, *13,* 445–452.

Rebattu, J., & Lafon, H. Les troubles de l'olfaction. *Le Journal de médicine de Lyon,* 1970, *51,* 2009–2020.

Reese, T. S. Olfactory cilia in the frog. *Journal of Cellular Biology,* 1965, *25,* 209–230.

Rehn, T. Perceived odor intensity as a function of airflow through the nose. *Sensory Processes,* 1978, *2,* 198–205.

Rehn, T. Reply to the comment on "Perceived odor intensity as a function of air flow through the nose." *Sensory Processes.* 1979, *3,* 286–288.

Rieser, J., Yonas, A., & Wikner, K. Radial localization of odors by newborns. *Child Development,* 1976, *47,* 856–859.

Riggs, L. A., Ratliff, F. C., Cornsweet, T., & Cornsweet, J. The disappearance of steadily fixated objects. *Journal of the Optical Society of America,* 1953, *43,* 495–501.

Romney, A. K., Shepard, R. N., & Nerlove, S. B. (Eds.). *Multidimensional scaling: Theory and application in the behavioral sciences* (2 vols.). New York: Seminar Press, 1972.

Rosen, A. A., Peter, J. B., & Middleton, E. M. Odor threshold of mixed organic chemicals. *Journal of Water Pollution Control Federation,* 1962, *7*–14.

Rosenbaum, J. B. The significance of the sense of smell in the transference. *Journal of the American Psychoanalytic Association,* 1961, *9,* 312–323.

Rovee, C. K., Cohen, R. Y., & Shlapack, W. Life span stability in olfactory sensitivity. *Developmental Psychology,* 1975, *11,* 311–318.

Rubert, S. L., Hollender, M. H., & Mehrhof, E. G. Olfactory hallucinations. *Archives of General Psychiatry,* 1961, *5,* 121–126.

Rummo, N. J., & Engen, T. *The effect of carbon monoxide and alcohol on odor detection.* Unpublished manuscript. Providence, Rhode Island: Injury Control Research Laboratory, 1973.

Russel, M. J. Human olfactory communication. *Nature (London),* 1976, *260,* 520–522.

Russel, M. J., & Mendelson, T. Mothers know their little stinkers. Personal communication.

Russel, M. J., Switz, G. M., & Thompson, K. Olfactory influences on the human menstrual cycle. *Biochemistry and Behavior,* in press.

Sarnat, H. B. *Olfactory reflexes in the newborn.* Paper presented at the First International Congress of Child Neurology, Toronto, October 7–10, 1975.

Schall, B., Montagner, H., Hertling, E., Bolzoni, D., Moyse, R., & Quichon, R. Les stimulations olfactives dans les relations entre l'enfant et la mère. *Reproduction, Nutrition, Development,* 1980, *20,* 843–858.

Schachtel, E. G. *Metamorphosis on the development of affect, perception, attention, and memory.* New York: Basic Books, 1959.

Schachter, S. Some extraordinary facts about obese humans and rats. *American Psychologist,* 1971, *26,* 129–144.

Schechter, P. J., & Henkin, R. I. Abnormalities of taste and smell after head trauma. *Journal of Neurology, Neurosurgery, and Psychiatry,* 1974, *37,* 802–810.

Schemper, T., Voss, S., & Cain, W. S. Odor identification in young and elderly persons: Sensory and cognitive limitations. *Journal of Gerontology,* 1980, *1,* 225–231.

Schiffman, S. S. Contributions to the physiochemical dimensions of odor: A psychophysical approach. *Annals of the New York Academy of Sciences,* 1974, *237,* 164–183. (a)

References

Schiffman, S. S. Physiochemical correlates of olfactory quality. *Science,* 1974, *185,* 112. (b)
Schiffman, S. S. Preference: A multidimensional concept. In J. H. A. Kroeze (Ed.), *Preference behaviour and chemoreception.* London: Information Retrieval Limited, 1979. Pp. 63–81.
Schiffman, S. S., Moss, J., & Erickson, R. P. Threshold of food odors in the elderly. *Experimental Aging Research,* 1976, *2,* 389–398.
Schiffman, S., Orlandi, M., & Erickson, R. P. Changes in taste and smell with age: Biological aspects. In J. M. Ordy & K. Brizzee (Eds.), *Sensory systems and communication in the elderly.* New York: Raven Press, 1979. Pp. 247–268.
Schiffman, S. S., Robinson, D. E., Erickson, R. P. Multidimensional scaling of odorants: Examination of psychological and physiochemical dimensions. *Chemical Senses and Flavor,* 1977, *2,* 375–390.
Schneider, R. A. Anosmia: Verification and etiologies. *Annals of Otology, Rhinology and Laryngology,* 1972, *81,* 272–277.
Schneider, R. A. Newer insights into the role and modifications of olfaction in man through clinical studies. *Annals of the New York Academy of Sciences,* 1974, *237,* 217–223.
Schneider, R. A., Costiloe, P., Howard, R., & Wolf, S. Olfactory perception thresholds in hypogonadal women: Changes accompanying the administration of androgen and estrogen. *Journal of Clinical Endocrinology and Metabolism,* 1958, *18,* 379–390.
Schneider, R. A., Costiloe, J. P., Vega, A., & Wolf, S. Olfactory threshold technique with nitrodilution of n-butane and gas chromatography. *Journal of Applied Physiology,* 1963, *18,* 414–417.
Schneider, R. A., & Schmidt, C. E. Dependency of olfactory localization on nonolfactory cues. *Physiology and Behavior,* 1967, *2,* 305–309.
Schneider, R. A., Wolf, S. Olfactory perception thresholds for citral utilizing a new type of olfactorium. *Journal of Applied Physiology,* 1955, *8,* 337–342.
Self, P. A., Horowitz, F. D., & Paden, L. Y. Olfaction in newborn infants. *Developmental Psychology,* 1972, *4,* 349–363.
Sellzer, R. T. Monsanto develops malodor counteractant. *Chemical Engineering News,* 1975, *53,* 24–25.
Shepard, R. N. Attention and the metric structure of the stimulus space. *Journal of Mathematical Psychology,* 1964, *1,* 54–87.
Semb, G. B. The detectability of the odor of butanol. *Perception and Psychophysics,* 1968, *4,* 335–340.
Sem-Jacobsen, C. W., Petersen, M. C., Dodge, H. W., Jr., Jacks, Q. D., Lazarte, J. A. & Holman, C. B. Electic activity of the olfactory bulb in man. *American Journal of Medical Science,* 1956, *232,* 243–251.
Shepard, R. N. Recognition memory for words, sentences, and pictures. *Journal of Verbal Learning and Verbal Behavior,* 1967, *6,* 156–163.
Shepherd, G. M. Central processing of olfactory signals. In D. Müller-Schwarze & M. M. Mozell (Eds.), *Chemical signals in vertebrates.* New York: Plenum Press, 1976.
Shepherd, G. M., Getchell, T. V., & Kauer, J. S. Analysis and function in the olfactory pathway. In D. B. Tower (Ed.), *The nervous system. The basic neurosciences.* New York: Raven Press, 1975.
Shigeta, Y. Research on odor abatement and control in U.S.A. (II) *Akusho no Kenkyu (Odor Research Journal of Japan),* 1971, *1,* 9–20. (In Japanese)
Siegel, S. The role of conditioning in drug tolerance and addition. In J. D. Keeln (Ed.), *Psychopathology in animals: Research and clinical applications.* New York: Academic Press, 1979.

Slosson, E. E. A lecture experiment in hallucinations. *Psychological Review,* 1899, *6,* 407–408.
Slotnick, B. M., & Ptak, J. E. Olfactory intensity—Difference thresholds in rats and humans. *Physiology and Behavior,* 1977, *19,* 795–802.
Smith, C. S. Age incidence of atrophy of olfactory nerves in man. *Journal of Comparative Neurology,* 1942, *77,* 589–595.
Sparkes, R. S., Simpson, R. W., & Paulsen, C. A. Familial hypogonadotropic hypogonadism with anosmia. *Archives of Internal Medicine,* 1968, *121,* 534–538.
Spence, W., & Guilford, J. P. The affective value of combinations of odors. *American Journal of Psychology,* 1933, *45,* 495–501.
Springer, K. J. Combustion odors—A Case study. In A. Turk, J. W. Johnston, Jr., & D. G. Moulton (Eds.), *Human responses to environment odors.* New York: Academic Press, 1974. Pp. 227–262.
Springer, K. J., & Stahman, R. C. Control of disease exhaust odors. *Annals of the New York Academy of Sciences,* 1974, *237,* 409–426.
Stein, M., Ottenberg, P., & Roulet, N. A study of the development of olfactory preferences. *American Medical Association Archives of Neurological Psychiatry,* 1958, *80,* 264–266.
Steiner, J. E. Facial expressions of the neonate indicating the hedonics of food-related chemical stimuli. In J. M. Weittenback (Ed.), *Taste and development.* Bethesda, Md.: DHEW Publication No. (NIH) 77-1068, 1977. Pp. 173–189.
Steinmetz, G., Pryor, G. T., & Stone, H. Olfactory adaptation and recovery in man as measured by two psychophysical techniques. *Perception and Psychophysics,* 1970, *8,* 327–330.
Sternberg, S. High-speed scanning in human memory. *Science,* 1966, *153,* 652–654.
Stevens, J. C., & Stevens, S. S. Brightness functions: Effects of adaptation. *Journal of the Optical Society of America,* 1963, *53,* 375–385.
Stevens, S. S. (Ed.). *Handbook of experimental psychology.* New York: Wiley, 1951. Pp. 1–49.
Stevens, S. S. *Psychophysics.* New York: Wiley, 1975.
Stone, H., & Bosley, J. J. Olfactory discrimination and Weber's law. *Perceptual and Motor Skills,* 1965, *20,* 657–665.
Stone, H., & Pangborn, R. M. Intercorrelation of the senses. *Basic Principles of Sensory Evaluation, Special Technical Publication No. 433.* Philadelphia: American Society for Testing and Materials, 1968.
Studdert-Kennedy, M., Lieberman, A. M., Harris, K. S., Cooper, F. S. Motor theory of speech perception. *Psychological Review,* 1970, *77,* 234–249.
Stuiver, M. *Biophysics of the sense of smell.* Doctoral dissertation, University of Grooningen, 1958.
Subcommittee on hydrogen sulfide. *Hydrogen sulfide.* Baltimore: University Park Press, 1979.
Sullivan, F., & Leonardos, G. Determination of odor sources for control. *Annals of the New York Academy of Sciences,* 1974, *237,* 339–349.
Sumner, D. Post-traumatic anosmia. *Brain,* 1964, *87,* 107–120.
Svensson, L. T., & Lindvall, T. On the consistence of intramodal intensity matching in olfaction. *Perception and Psychophysics,* 1974, *16,* 264–270.
Tanabe, T., Iino, M., & Takagi, S. F. Discrimination of odors in olfactory bulb, pyriform–amygdaloid areas, and orbitofrontal cortex of the monkey. *Journal of Neurophysiology,* 1975, *38,* 1284–1296.
Tanabe, T., Yarita, H., Iino, M., Ooshima, Y., & Tagaki, S. F. An olfactory projection area in orbitofrontal cortex of the monkey. *Journal of Neurophysiology,* 1975, *38,* 1269–1283.

Teghtsoonian, M., Teghtsoonian, R., Berglund, B., & Berglund, U. Comment on "Perceived odor intensity as a function of airflow through the nose." *Sensory Processes*, 1979, *3*, 204–206.

Teghtsoonian, R., Teghtsoonian, M., Berglund, B., & Berglund, U. Invariance of odor strength with sniff vigor: An olfactory analogue to size constancy. *Journal of Experimental Psychology: Human Perception and Performance*, 1978, *4*, 144–152.

Third Karolinska Symposium on Environmental Health. *Nordisk Hygienisk Tidskrift*, 1970, *51*, 1–77.

Thomas, L. *The Lives of a Cell*. New York: Viking, 1974.

Torda, C., & Wolff, J. G. Effect of steroid substances on synthesis of acetylcholine. *Proceedings of the Society of Experimental and Biological Medicine*, 1944, *57*, 327–390.

Tucker, D. Physical variables in the olfactory stimulation process. *Journal of General Physiology*, 1963, *46*, 453–489.

Tucker, D. Nonolfactory responses from the nasal cavity: Jacobson's organ and the trigeminal system. In L. M. Beidler (Ed.), *Handbook of sensory physiology*. Vol. IV. *Chemical senses 1: Olfaction*. New York: Springer-Verlag, 1971. Pp. 151–181.

Turk, A. Approaches to sensory measurement. *Annals of the New York Academy of Sciences*, 1964, *116*, 564–566.

Uttal, W. R. *The psychophysiology of sensory coding*. New York: Harper & Row, 1973.

Van Boxtel, A., & Köster, E. P. Adaptation of the electro-olfactogram in the frog. *Chemical Senses and Flavor*, 1978, *3*, 39–44.

Vaschide, N. Researches experimentales sur la fatigue olfactive. *Journal de L'Anatomie et de la Physiologie*, 1902, *38*, 85–103.

Venstrom, D., & Amoore, J. E. Olfactory threshold in relation to age, sex or smoking. *Journal of Food Science*, 1968, *33*, 264–265.

von Fiandt, K. *The world of perception*. Homewood, Ill.: Dorsey Press, 1966.

von Hornbostel, E. M. The unity of the senses. *Psyche*, 1927, *7*, 83–89.

von Hornbostel, E. M. Die Einheit der Sinne. English translation in W. D. Ellis (Ed.), *Source book of gestalt psychology*. London: Routledge and Kegan Paul, 1938.

von Sydow, E. Comparison between psychophysical and electrophysiological data for some odor substances. In N. Tanyolac (Ed.), *Theories of odors and odor measurement*. Bebek, Istanbul: Robert College Research Center, 1968. Pp. 297–301.

Warren, R. M., & Warren, R. P. Auditory illusions and confusions. *Scientific American*, December, 1970, *223*, 30–36.

Wells, F. L. Reaction-times to affects accompanying smell stimuli. *American Journal of Psychology*, 1929, *41*, 83–86.

Wender, I. Intensität und qualität in der Geruchs Wahrnehmung. *Psychologische Forschung*, 1968, *32*, 244–276.

Wenger, M. A., Jones, F. N., & Jones, M. H. *Physiological psychology*. New York: Holt, Rinehart, and Winston, 1956.

Wenzel, B. M. Techniques in olfactometry: A critical review of the last one hundred years. *Psychological Bulletin*, 1948, *45*, 231–247.

Wenzel, B. M. Differential sensitivity in olfaction. *Journal of Experimental Psychology*, 1949, *39*, 124–143.

Wheeler, R. H. The synesthesia of a blind subject. *University of Oregon Publications*, 1920, *1*, 1–61.

Whitten, W. K. Modification of the oestrus cycle of the mouse by external stimuli associated with the male. *Journal of Endocrinology*, 1956, *13*, 399–404.

Wiener, H. External chemical messengers. I. Emission and reception in man. *New York Journal of Medicine*, 1966, *66*, 3153–3170.

Wiener, H. External chemical messengers. II. Natural history of schizophrenia. *New York State Journal of Medicine,* 1967, *67,* 1144–1165. (a)
Wiener, H. External chemical messengers. III. Mind and body in schizophrenia. *New York State Journal of Medicine,* 1967, *67,* 1287–1310. (b)
Wiener, H. External chemical messengers. IV. Pineal gland. *New York State Journal of Medicine,* 1968, *68,* 912–938. (a)
Wiener, H. External chemical messengers. V. More functions of the pineal gland. *New York State Journal of Medicine,* 1968, *68,* 1019–1038. (b)
Woodrow, H. F., & Karpman, B. A new olfactometric technique and some results. *Journal of Experimental Psychology,* 1917, *3,* 431–447.
Woodworth, R. S., & Schlosberg, H. *Experimental psychology.* New York: Holt, 1954.
Woskow, M. H. Multidimensional scaling of odors. In N. Tanyolac (Ed.), *Theories of odors and odor measurement.* Bebek, Istanbul: Robert College Research Center, 1968. Pp. 147–191.
Wotman, S., Mandel, I. D., Khotim, S., Thompson, R. H., Kutscher, A. H., Zegarelli, E. V., & Denning, C. R. Salt thresholds and cystic fibrosis. *American Journal of Disease in Children,* 1964, *108,* 372–374.
Wright, R. H. *The science of smell.* New York: Basic Books, 1964.
Wright, R. H. Why is an odour? *Nature,* 1966, *209,* 551–554.
Wright, R. H. Predicting olfactory quality from far infrared spectra. *Annals of the New York Academy of Sciences,* 1974, *237,* 129–136.
Wright, R. H. Odor and molecular vibration: Neural coding of olfactory information. *Journal of Theoretical Biology,* 1977, *64,* 473–502.
Wright, R. H. The perception of odor intensity: Physics or psychophysics II. *Chemical Senses and Flavor,* 1978, *3,* 242–245. (a)
Wright, R. H. Specific anosmia: A clue to the olfactory code or to something much more important? *Chemical Senses and Flavor,* 1978, *3,* 235–239. (b)
Wysocki, C. J. Neurobehavioral evidence for the involvement of the vomeronasal system in mammalian reproduction. *Neuroscience and Biohavioral Reviews,* 1979, *3,* 301–341.
Yamazaki, K., Yamaguchi, M., Baronoski, L., Bard, J., Boyse, E. A., & Thomas, L. Recognition among mice: Evidence from the use of a Y-maze differentially scented by congenic mice of different major histocompatibility types. *Journal of Experimental Medicine,* 1979, *150,* 755–760.
Yoshida, M. Studies of psychometric classification of odors (4). *Japanese Psychological Research,* 1964, *6,* 115–124. (a)
Yoshida, M. Studies of psychometric classification of odors. (5). *Japanese Psychological Research,* 1964, *6,* 145–154. (b)
Yoshida, M. In search of the fundamental scheme of olfaction. In Y. Katsuki, M. Sato, S. F. Takagi, & Y. Oomura. *Food intake and the chemical senses.* Tokyo: University of Tokyo Press, 1977. Pp. 113–128.
Yoshida, M. Descriptive and emotional profiles of odours and their preferences. In J. H. A. Kroeze (Ed.), *Preference behaviour and chemoreception.* London: Information Retrieval Limited, 1979. Pp. 83–92.
Youngentob, S. L., Kurtz, D. B., Leopold, D. A., Mozell, M. M., & Horning, D. E. Olfactory sensitivity: Is there laterality? *Sensory Processes,* in press.
Zigler, M. J., & Holoway, A. H. Differential sensitivity as determined by the amount of olfactory substance. *Journal of General Psychology,* 1935, *12,* 372–382.
Zwaardemaker, H. *Die Physiologie des Geruchs.* Leipzig: Engelmann, 1895.
Zwaardemaker, H. Die compensation von Geruchsemphindungen. *Archives für Anatomie und Physiologie* [*Physiological Abstracts*], 1900, 423–432.
(See *Perfumery and Essential Oil Review,* 1959, *50,* 217–221.)

References

Zwaardemaker, H. *L'Odorat*. Paris: Doin, 1925.
Zwaardemaker, H. An intellectual history of a physiologist with psychological aspirations. In C. Murchison (Ed.), *A history of psychology in autobiography* Vol. 1. New York: Russell and Russell, 1961.

SUBJECT INDEX

A

Absolute threshold, *see* Threshold
Acoustic neuroma, 150
Acuity, 3–5, *see also* Sensation
Adaptation, 61–74
   asymetric effect in cross-adaptation, 119
   classic findings, 15–16
   cross-adaptation, 74, 171
   effect of duration of adapting condition, 68
   effect of intensity of adapting condition, 74
   function of number of olfactory receptors, 22
   recovery from, 74
   versus habituation, 171
Addison's disease, 92
Adequate stimulus, 151
Age and preference in children, 135
Aging and odor perception, 90–92
Air pollution
   effect on odor perception, 77
   mixture of pollutants and odor intensity, 113
   odor-control technology, 42
   psychophysics of odors and pollutants, 163
   role of odors in air pollution, 126
Albino, effect on odor perception, 81

Aliphatic acids, 141
Alliesthesia, 77, 138, 154
Allocentric perception, 129
Amygdaloid complex, 23, 28
Analytic odor perception, 123
Androgonadic stimulation, 82
Androstenol, 141, 142, 144
Animal behavior, 4, 13
Anorexia nervosa, 94, 111, 165
Anosmia, 1–2, 79–95
   causes, 87–89
   effect of drugs on, 89, 91
   effect of pollutants on, 89–91
   effect on appetite, 94–95
   effect on arousal, 95
   effect on sex drive, 93–95
   eunuchoidism, 82
   functional, by preventing olfactory stimulation, 143, 146
   general, 81
   head injury, 87
   hydrogen cyanide gas, 81
   hypogonadotrophic hypogonadism, 82, 95
   Kallmann's syndrome, 81
   sex-linked, 82
   specific or partial, 30, 83
Anterior olfactory nucleus, 23
Aroma, 145

Arousal, 80, 127, 153–154, 155, 171
Artificial nose, 59, 164–165, see also
  Physical sensors
Association to odor perception, 97, 127
  acquired equivalence of olfaction and
    other modalities, 155
  aversive therapy with unpleasant odors,
    110, 126, 165
  biologically significant odors, 169, 170
  effect on preference, 97, 127, 143
  food aversion, 165
  odor as a nonfunctional component, 130
Atrophy of olfactory receptors
  aging, 90
  anosmia, 89
  training of odor perception, 161
Audition, effect on odor perception, 153
Autocentric perception, 129

B

Bait shyness, 14, 110
Biologically significant odors, 144, 169, 170
Blast-injection olfactometer, 37
Body odor, 139, 143–144
  racial differences, 13
Bombykol, 125, 142
Bombyx mori, 125, 140
Bruce effect, 166

C

Cacosmia, 86, 87
Camera inodorata, 6
Categorical perception, 6
Child development in odor preference,
  130–134, see also Infant responses to
  odors
Coding
  effect of cognitive factors, 49
  history, 10
  insects, 25
  models for olfaction, 172
  olfactory bulb, 27
  receptor theories of odor intensity and
    quality, 29–33
Cognitive factors in odor perception, 2,
  129, 155
Compensation, 9, 77, 114, 115, 116
Compromise, 116
Conditioning, 165
Constant-sensation method, 67

Context effect, 159
Copulin, 141
Cribriform plate, 26, 88
Cross-adaptation, 74–76, 171
Cross-fiber pattern, 33
Cystic fibrosis, 92

D

Deodorizer
  adaptation, 118
  effect of intensity, 116
  mixture of odors, 114
  psychophysics, 116
  role in air pollution, 126
Dichorionic stimulation, 116, 120, 152
Diesel fumes, 136
Diet, 165
  aversion therapy with unpleasant odors,
    110, 126, 156
Difference threshold, 54–56
  Weber's Law, 55
Discrimination of odors
  difference threshold, 54–56
  distinguished from odor recognition, 100
  odor quality versus intensity, 101, 115
Drugs, effect on odor perception, 91
Dysosmia, 94

E

ECM, see External chemical messengers
Electro-olfactogram (EOG), 23, 24, 25, 29,
  73, 74, 115
Emission stacks, 43
Emotion, relation with olfaction, 109
Enhancer of odor perception, 153, see also
  Facilitation
EOG, see Electro-olfactogram
Equilibrium, sniffer, 38
Estrogen, effect on odor sensitivity, 83, 143
Eunuchoidism, 82
Exaltolide, 144
Excitation of olfactory receptors, 22–24,
  27, 30
External chemical messengers (ECM), 140

F

Facilitation
  neurological, 152
  psychophysical, 76, 115
False alarm in odor detection, 158, see
  also Mental set; Response bias

Subject Index

Filia olfactoria, 23
Flavor, 145, 147
Flow rate of odor stimulus, 19, 160
Food aversions, 14, 110–111, 170–171
Forgetting of odors, 110
Free nerve endings, 21
Fresh air smell, 136

G
Gas chromatographic model, 33–34, 144
   masking and adaptation, 119
Glomerulus, 23, 23–25
Gonadotrophic action, 166
Gustation, 145–149
   without odor perception, 145, 147

H
Habituation of odor response, 62–63, 171
Hallucination of odors, 86, 87, 89, 158
Health, effect of odor, 137–170
Hedonics, 11–14, 126, 136, 137
Henning, Hans, 8, 116
Hereditary deficiency in odor perception, 80–86
Histocompatibility complex, 13, 144
Homeostasis, 138
Hormone effect on odor sensitivity, 166, see also Menstrual cycle and odor sensitivity
Human sensitivity, compared with physical sensors, 59
Hyperosmia, 86–87, 92–93
   Addison's disease
   biologically significant odorant, 93
   cystic fibrosis, 92
   menstrual cycle, 93
   relative hyperosmia, 93
Hypogeusia, 16, 94
Hypogonadism, 82
Hypogonadotrophic hypogonadism, 82, 95
Hyposmia, 84, 86, 89
   menstruation, 93, 140
Hypothalamus, 2, 3, 28, 82
Hysteria, 89

I
Ideopathic hypogeusia, 94
Illusions of odors, 157–159
Indoor air quality, 126
Infant responses to odors
   adaptation, 62

   diagnostic value, 80
   facilitation, 76
   preference, 13, 130
Information theory, 92, 161
Inhibition of neural response, 26–27, 28, 30, 152
Irritation, see Trigeminal nerve

J
Jacobson's organ (vomeronasal organ), 21

K
Kallmann's syndrome, 82, 95
Korsakoff's syndrome, 109

L
Lateralization of olfactory function in brain, 17
Light, interaction with odor, 153
Linnaeus, 45
Localization of odorants in environment, 59
Long-term memory, 107

M
Malodor, 134–135, 137, 163
Masking, 44, 116
   adaptation, 118
Memory for odors, 14, 97, see also Association to odor perception; Food aversions
   sex attraction, 143, 170
Menopause, 91
Menstrual cycle and odor sensitivity, 93, 140, 144
Mental set
   response bias, 149
   verbal labels, 149
Methathetic dimension, 123
Method of absolute judgment, 102
Method of magnitude estimation, 57, 59
Microencapsulation of odorants, 38, 40
Mitral cells, 23, 26, 27
Mixture of odors, 113
   dichorionic, 152
   odorants and tastants, 147, 149
   psychophysical principles, 116
   vector model of odor mixture, 120–123
Molecular vibration theory, 32
Multidimensional scaling of odors, 32, 45, 47–51, 126
Musk, 167, 170

N

Naming odors, 14, 102–103
  intensity versus quality, 101
  with versus without labels, 102, 106
Naris, 17–19
Nasal passage, see Naris
Nervus terminalis, 22
Noise, effect on detection of odor, 154
Nostril, see Naris
Nutrition, effect on odor sensitivity, 169

O

Obesity, 145
Odorant, definition, 2
Odor classification, 7–9, 44–47
  influence of verbal labels, 102
  stereochemical theory, 31
Odor control technology, 42–44
Odor definition, 2
Odor experts, 4–5, 101–102
Odor identification, 98–99
Odor intensity, compared with odor quality, 101, 115, 120
Odor labels, 103–106
Odor memory, 14, 97
  sex attraction, 143, 170
Odor mixture, see Mixture of odors
Odor preference
  air pollution, 137
  association, 143–144
  bodily state or need, 138, 139
  complexity of odorant, 136
  effect of learning experience, 169
  familiarity, 127
  individual differences, 127, 139
  nonfunctional role of odor, 169, 171
  opinion poll, 137
  origin, development through experience, 13, 130
  repeated exposure of odor, 139
  response bias, 13
  sex difference, 135
Odor quality, compared with odor intensity, 101, 115, 120
Odor recall, 111–112, 153
Odor recognition, 100, see also Odor memory
Odor repellents, 130, 139
Odor sensitivity, see also Threshold
  aging, 90–92
  alcohol, 91
  cocaine, 91
  females versus males, 144
  medical examination, 80
Odor similarity, 47–51, 109
Odor stimulus, 6–7, 35–36
Olfactie, 70
Olfactometry, 36–42
Olfactory arousal, 95
Olfactory brain, 27–29
Olfactory bulb, 23, 25–27
Olfactory cilia, 21, 23
Olfactory cleft, 18
Olfactory deficiencies, 79–95
  classification, 86–89
Olfactory epithelium, 18–19, 20, 23
  effect of vitamin A, 89
Olfactory knob, 20
Olfactory mucosa, 19, 20, 23, 34
Olfactory nerve, 23
Olfactory receptors, 70–25
  atrophy, 89, 90
  generalists, 25
  regeneration, 90
  as single units, 27
  specialists, 2
Olfactory slit, 18
Olfactory tract, 23, 26
Olfactory tubercle, 23, 28
Ovariectomy, 83, 142
Overstimulation, 151
Ovulation, 140, 144

P

Parosmia, 86–87
Pattern theory, 33
  cross-fiber pattern, 33
  gas chromatographic model, 33
Penetrations-and-puncturing theory, 31–32
Perception, 3
Perceptual constancy, 19, 157, 159–161
Perfumers, ability to judge odors, 101–102
Perfumes
  advertising, 130
  age, 135
  culture, 134
  gender, 135
  marketing, 167
  musk, 167
Periglomerular cells, 25

Phantasmia, 89
Pheromone, 125
 bombykol, 125
 external chemical messengers, 140
 human, 139
 learning hypothesis, 169
 vomeronasal system or Jacobson's organ, 21–22
Physical sensors, compared with human olfactory ability, 4, 59, 137
Potentiation, 113, see also Facilitation; Synergism
Power function, see Psychophysical power function
Preference, see Odor preference
Pregnancy block, 166
Prepyriform cortex, 23, 28
Primary odorant, 27
Proactive interference (or inhibition), 110, 111, 170
Processing of odor information, serial and parallel, 156
Prothetic dimension, 123
Proust, 14, 15, 98–99
Psychophysical power function, 57, 58, 122
Psychophysical scaling, 56–58
Psychophysics, 5, 15, 35
 indoor air quality, 126
 mixture, 120–123
 relevance to definition of pheromone, 160
 scaling odor annoyance

R
Raman spectrum, 32
Recall, see Odor recall
Recognition, see Odor recognition
Recovery of adaptation, 74
Regeneration of olfactory nerves, 90
Repellents, see Odor repellents
Repression of odor experience, 12
Respiratory disturbances, 86
Response bias, see also Mental set
 effect of expectation in judging flavor, 146
 false alarms, 158
 opinion survey on children's preference, 131
 signal detection theory, 52
Retroactive interference, 110, 111, 170

S
Scrubbers, 42
Self-adaptation, 63, see also Adaptation
Semantic memory, 102–103
Sensation, 2
Sensor, see Artificial nose; Physical sensors
Sex attractants, 141–143, see also Pheromone
Sex difference in odor preference, 135
Sexual development and olfaction, 95
Short-term memory, 107, 109
Signal-detection theory, 52–54
Similarity, see Odor similarity
Smoking, 75
Sniff bottles, 38
Sniffmobile, 40–41
Sniff strips, 38
Sound, interaction with odor, 153
Stereochemical odor classification theory, 30–31, 45
 specific anosmia, 83, 85
Sweat, 141, 143, 167
 body odor, 143
 effect on perfume, 167
Synergism, 114, 115, see also Facilitation; Potentiation
Synesthesia, 153

T
Tampon method, 152, see also Vaginal odor
Taste and odor interaction, 145–149
Taste-blindness, 16
Testosterone, 83, 141
Threshold, 51–52, see also Odor sensitivity
 change through menstrual cycle, 140
 contrasted with signal-detection theory, 52–54
 variability with heredity, 81
Tip-of-the-nose phenomenon, 50, 102
Tobacco smoke, 136
Training of odor perception, 161
Transitional entorhinal cortex, 23
Trigeminal nerve, 2
 acoustic neuroma, 150
 contribution to food identification, 147
 contribution to odor intensity, 21, 150
 inhibition of respiration, 151
 irritation and odor confused, 149, 151
 response by human infants, 132
Tufted cells, 23, 26–27

U

Unity of the senses, 155, 172
Urine, 141

V

Vaginal odors, 143
Vector model of odor mixture, 120
  taste-odor interaction, 149
Verbal labels, influence on odor memory, 102
Visual stimulation, effect on odor perception, 153
Vitamin A, treatment of anosmia, 89
Vomeronasal organ (Jacobson's organ), 21, 28

W

Weber's Law, 55

Z

Zwaardemaker, 3, 9, 36, 70, 116

ACADEMIC PRESS
SERIES IN COGNITION AND PERCEPTION

SERIES EDITORS:
Edward C. Carterette
Morton P. Friedman
Department of Psychology
University of California, Los Angeles
Los Angeles, California

Stephen K. Reed: *Psychological Processes in Pattern Recognition*

Earl B. Hunt: *Artificial Intelligence*

James P. Egan: *Signal Detection Theory and ROC Analysis*

Martin F. Kaplan and Steven Schwartz (Eds.): *Human Judgment and Decision Processes*

Myron L. Braunstein: *Depth Perception Through Motion*

R. Plomp: *Aspects of Tone Sensation*

Martin F. Kaplan and Steven Schwartz (Eds.): *Human Judgment and Decision Processes in Applied Settings*

Bikkar S. Randhawa and William E. Coffman: *Visual Learning, Thinking, and Communication*

Robert B. Welch: *Perceptual Modification: Adapting to Altered Sensory Environments*

Lawrence E. Marks: *The Unity of the Senses: Interrelations among the Modalities*

Michele A. Wittig and Anne C. Petersen (Eds.): *Sex-Related Differences in Cognitive Functioning: Developmental Issues*

Douglas Vickers: *Decision Processes in Visual Perception*

Margaret A. Hagen (Ed.): *The Perception of Pictures, Vol. 1: Alberti's Window: The Projective Model of Pictorial Information, Vol. 2 Dürer's Devices: Beyond the Projective Model of Pictures*

Graham Davies, Hadyn Ellis and John Shepherd (Eds.): *Perceiving and Remembering Faces*

Hubert Dolezal: *Living in a World Transformed: Perceptual and Performatory Adaptation to Visual Distortion*

Gerald H. Jacobs: *Comparative Color Vision*

Diana Deutsch (Ed.): *The Psychology of Music*

John A. Swets and Ronald M. Pickett: *Evaluation of Diagnostic Systems: Methods from Signal Detection Theory*

Trygg Engen: *The Perception of Odors*

in preparation

C. Richard Puff (Ed.): *Handbook of Research Methods in Human Memory and Cognition*